Behind Time

The incoherence of time and McTaggart's atemporal replacement

GERALD ROCHELLE

Routledge
Taylor & Francis Group
LONDON AND NEW YORK

First published 1998 by Ashgate Publishing

Reissued 2018 by Routledge
2 Park Square, Milton Park, Abingdon, Oxon, OX14 4RN
711 Third Avenue, New York, NY 10017, USA

Routledge is an imprint of the Taylor & Francis Group, an informa business

A Library of Congress record exists under LC control number: 97077891

ISBN 13: 978-1-138-61020-0 (hbk)
ISBN 13: 978-0-429-46069-2 (ebk)

Contents

Preface

I am indebted to Dr Keith Hossack at King's College London for his help and encouragement without which this project never would have been completed. Also to Dr Tim Crane at University College London for his important criticisms and comments. Thanks also to Dr Bill Mander at Harris Manchester College, Oxford for raising many points most of which I have tried to answer or accommodate. In addition, I am immeasurably grateful to Professor John Heywood Thomas who introduced me to McTaggart and who has patiently helped and guided me with detailed criticism and valuable amendments of this and all my other work on McTaggart. Lastly, I must acknowledge the support I have received over the years from Edward Senior of Cincinnati, Ohio, dear friend of McTaggart's most lucid expositor Stanley Victor Keeling, who, though having no direct part in the completion of this book, has had a great and beneficial influence over me and has not only kept my sometimes sinking enthusiasm afloat but has also encouraged me to write and express myself more clearly and efficiently.

1 Introduction

Time is puzzling. This work is an attempt to solve part of the puzzle by looking at McTaggart's argument against its reality and his vision of a replacement atemporal world. My opinion is that fully to understand McTaggart's view about time we need to go beyond an analysis of his denial that it is real. Many philosophers have become so involved in his argument against the reality of time that they fail to notice that it makes little difference whether the argument is right or wrong. If it is right then time must be something other than what it appears to be. If it is wrong then the difficulty in showing that it is wrong makes any assertion about the consequent version of time incompatible with time as we know it. After all, if the reality of time was self evident there would be no need to get involved in finding a defence to any argument against it.

McTaggart takes an idealist view that reduces all that is real to spirit; that any expression of reality is dependent on or a manifestation of consciousness. This is an interesting perception of the world which allies to the contemporary scientific world-view which holds a scepticism of apparent materiality. Whereas McTaggart's denunciation of time is well known his replacement for time, the *C* series, is not. McTaggart's system of atemporality forms part of a coherent metaphysical scheme that should be taken seriously. The system helps us to interpret some riddles that time presents to us. It also explains how we can form part of an ideal world that is not out of place in modern times.

Broadly, McTaggart's argument against time has two parts. Part one involves three steps. Step 1. Time involves change. Step 2. Change can only occur if there is a series of relations past, present and future, the *A* series. Step 3. The *A* series involves a contradiction that leads to a vicious infinite regress therefore there can be no time. Part two argues that the relation earlier and later, the *B* series, must be temporal, but that there can be no *B* series without an *A* series. As the *A* series is self-contradictory there can be no *B* series. This leads to two problems. Problem 1. Our view of things changing in the *A* series is wrong. Problem 2. Our experience of time in the *B* series is also wrong. These are problems because nothing seems more certain than our experience that things are changing in time.

Many philosophers accept that time requires change but conclude that there is no need to get involved with the *A* series because this is not essential for time. They think that for time to exist we need only the 'movement' from earlier to later (the *B* series) and as this relation is undeniably in place the argument that the *A* series is self-contradictory is irrelevant. This is the dominant *B*-theorist view which invites us to accept a temporal world that scarcely seems like time as we experience it. Some move closer to our everyday conception of time and accept the reality of the *A* series but hold that McTaggart's argument for its contradictory nature is spurious. This is the slightly less popular *A*-theorist view which has less philosophical appeal but is more in keeping with common sense. I think both these views (taken in a general sense) are misleading. On the first account we are left to deal with a form of time which does not seem to fit with our everyday experience and on the second we are left wondering that if time in the *A* series is not puzzling then why are we so puzzled about it.

I think we have three alternatives. 1. Accept the argument and look at a world without time. 2. Deny the argument and take one of two routes. *a*. Accept the reality of time and face unremitting *A* series problems, answers to which lead to a view of time that scarcely resembles how we think of it. *b*. Adopt an interpretation of time in the *B* series that scarcely resembles how we experience it. In either case still we are looking at a world without time. 3. Keep arguing about the argument. After ninety years this has proved inconclusive. If eventually there is a finding still we will be faced with either option '1' or '2' and as the outcome of each is substantially the same, we are not disadvantaged by considering a world without time.

My view is that McTaggart is correct in both the basis for his argument and his conclusion that time is unreal. No *B*-theorist argument gives an acceptable account of change. *A*-theorist arguments against the contradictory nature of the *A* series are very wire drawn and, though accepting the reality of time, dissolve the *A* series into something unlike how we think of the passage from future to present to past.

The nature of time in our experience is so uncertain that it makes either conclusion a possibility. However, making a broadly interpretable quality like time fit our purposes does not bring us any closer to its true nature. If we accept that time is very confusing, indeed so confusing as to raise a high level of suspicion about its reality (or at least it's coherence) in our minds, then we have good reason to look for a replacement. It is indeed strange that though *A* and *B*-theorists are prepared to create elaborate versions of time in order to defend it against McTaggart's denial, they are reluctant to look at anything atemporal that replaces it, even though such a replacement may be more 'time-like' than some of their own temporal variants. McTaggart offers a replacement by providing an atemporal scheme which can be substituted for what we think is time. This replacement system answers many of our confusions about temporality as well as pointing us towards answers about other confusions that appearance in general presents us with.

Here, I try to show that McTaggart's atemporal vision of reality, which gets little philosophical attention, is a serious attempt to describe a coherent world without time. To reinforce the need for us to consider this I have concentrated on showing that no matter how we look at time it is never acceptably coherent. I suggest that the answer to the puzzling nature of time is not to be found in its variable and confusing components but in the atemporal *C*-world that McTaggart describes.

Like McTaggart, I hold an idealist view, though perhaps a more 'absolutist' one than can be ascribed to McTaggart. I believe spiritual selfhood is the expression of the universe's ability to know itself. I think that consciousness is a special universal circumstance that has priority over the apparently objective world. It seems to me sensible that the conscious world can at least entertain this (whether it is right or wrong about what is true) whereas it seems absurd that the apparently objective world can be anything other than an object of consciousness. If, as I believe, this is so, then all apparently material aspects of the universe are subordinate to consciousness.

It is important that we know something of McTaggart before we begin. He is a complex philosopher whose devotion to metaphysical enquiry led him to relentless scrutiny of what could be the nature of reality.[1] In Chapter 2 I provide some initial information on McTaggart, his terms and his system. In § 2.1 I provide a brief synopsis of his important publications. In § 2.2 I give an explanation of some of his important terms. In § 2.3, as well as an outline of his whole system, I give a summary of the main features of his fuller philosophy, without which it is not possible to appreciate his atemporal system. In § 2.4 I conclude that McTaggart's argument against time is part of his fuller philosophy and is not given justice by being dealt with discretely.

In Chapter 3 I describe McTaggart's argument against time in more detail. In doing this I show that *B*-theory answers to McTaggart's claim that time is unreal leave the *B*-theorist denying time as we know it, in much the same way as McTaggart. This means that although *B*-theory arguments against McTaggart's position may have strength, they do little to move forward from a position of temporal scepticism. In § 3.2 (§§ 3.2.1-3.2.3) I explain McTaggart's argument against the reality of time. This leads to some problems in respect of how we think we experience time (§§ 3.2.4-3.2.5). In § 3.3 I show that McTaggart's argument also requires that we assume that it is events that change and not things. This assumption is difficult to accept and is almost universally rejected. In § 3.4 I look more closely at the vicious regress involved in the contradictory nature (according to McTaggart) of the phases of change in the *A* series. If the vicious regress can be successfully defended then McTaggart's argument would be greatly strengthened. To see if this is possible I analyse six arguments against the vicious regress. In § 3.5 I conclude that it is difficult to accept that it is events and not things that change. We need to take a strong idealist view to accept this. Even so, defence of the vicious regress can be found. McTaggart certainly provides us with enough suspicion about the reality of time to encourage us to see what it is he replaces it with.

In Chapter 4 I set out to see whether any of our commonly held views on time can be fully coherent. In § 4.2 I look at what I consider all the major views that either influence our opinions about time or to which we make recourse when we need some answer about its nature. In § 4.3 I look closely at the nature of the *A* series. In § 4.4 I conclude that none of our commonly held views provides a satisfactory answer. Indeed, having looked more closely at the *A* series (which, based on McTaggart's view, is essential for the reality of time) the matter only seems to get worse. It may be, however, that analysis of these views is misleading. It is possible that time will not yield to this sort of scrutiny and this, by itself, does not mean that it is unreal. However, I conclude that it is unsatisfactory that no conventional view provides an answer to some natural questions we have about time's nature.

Because the results in Chapter 4 are unsatisfactory, in Chapter 5 I look at some general positions on time that can be adopted in relation to our experience of it. My hope here is that since precise answers to time's puzzling nature may not be implicit in the way we describe it (our views of it), our experience of it (and positions we adopt to it) may provide something intuitively more positive. I look at three important positions that we can adopt when we consider time. In § 5.2 I look at the idea that there is 'reality of consciousness plus being': this is where we commonly expect to find an idealist like McTaggart. In § 5.3 I look at the idea that there is 'reality of becoming plus consciousness'. In § 5.4 I look at the idea held by science of 'a four-dimensional universe with no becoming'. In § 5.5 I conclude that McTaggart's position is surprisingly close to the views held by science. It also appeals to some of our common sense values. For these reasons, together with the conclusions arrived at in Chapter 4, in § 5.6, I justify why I think it is important to find out what he has to say on the subject.

In Chapter 6 I look at McTaggart's suggestion for an atemporal world as a replacement for the misleading world of time. In § 6.2 I describe

McTaggart's atemporal *C* series. In § 6.3 I describe how the atemporal series is coherent with McTaggart's fuller description of reality. In § 6.4 I describe how the *C* series can illuminate our puzzled view of time. Its 'inclusiveness' is similar to the apparent asymmetry of time (§ 6.4.1) and its 'direction' is similar to the apparent direction of the *B* series. In § 6.5 I describe some fundamental relations between the temporal world as we experience it and the atemporal *C* series. In § 6.6 I look more closely at the distinction between apparent and actual time. 'Apparent time' is the time in our experience that does not seem to resemble what time may really be (many of our views on time are like this). 'Actual time' is the measurable time of our everyday world. Actual time and apparent time I call '*true*' time and the atemporal order that in McTaggart's idealist world is reality I call '*really true*' 'time'. In this way we can keep the confusing notion of our experience of actual time as *true* time, separate from a plausible explanation of what it may really be in *really true* 'time'. In § 6.7 I conclude that it is reasonable to look beyond McTaggart's claim that time is unreal. However, his atemporal world is not easy to accept in total. Indeed, this is at the root of the problem. Because he proposes a complete metaphysics we get easily bound up with his ontological quest and may find ourselves rejecting atemporality for the wrong reasons.

In my Conclusion I provide a summary of the conclusions and explain why I think pursuing the idea of atemporality is worthwhile. I propose that although McTaggart's 'dynamic' view demands much of us this does not necessarily lead us to wholesale rejection of his thesis. I also conclude that his personal preoccupation with the loving self may (for some) detract from the logic of his argument. Nevertheless, I find that there is something in his argument for the atemporal self that has psychological appeal. Overall, I conclude that some elements of his atemporal vision are worthy of further scrutiny in the contemporary world. They should not be thrown out just because they form part of a rather suspicious whole. I conclude that we can accept the atemporal features of his argument without having to accept his system in total. The central elements of the atemporal world have psychological appeal and this idealist view is not out of place in modern times. I argue that McTaggart's atemporal world is a strong foundation for a new atemporal thinking which can at last break away from the confusing philosophical models of temporality that confuse us more than time itself.

Note

1 G. Rochelle, *The Life and Philosophy of J.McT.E. McTaggart, 1866-1925* (Lewiston, New York, Edwin Mellen, 1991).

2 McTaggart's Philosophy

2.1 McTaggart's works

In this section I provide a brief synopsis of McTaggart's important publications.

McTaggart's full metaphysical exposition is found in *The Nature of Existence*.[1] A popular version of the main course of his thinking can be found in *Some Dogmas of Religion*.[2] His important essays are collected in *Philosophical Studies*.[3]

Philosophical Studies contains the essence of his idealist philosophy. All the fundamental aspects of metaphysical enquiry are tackled: the existence of God, belief and the nature of mysticism,[4] time,[5] eternity,[6] causality,[7] the nature of self,[8] and the central, and connected issues of immortality, the nature of good, individual purpose and value.[9] There is also a brief clarification on the question of propositions applicable to themselves,[10] his reasons for considering philosophy of prime importance,[11] and an outline of how he thought metaphysics best defined.[12] An overview of McTaggart's intention is best had from his essay 'An Ontological Idealism' published in 1924. A comparison of this with one of his earliest works, 'The Further Determination of the Absolute' (written in 1893), reveals the remarkable consistency of his conclusions. It also shows the change from the earlier, Hegelian style of dialectical argument[13] obviously influenced by Bradley in its content though not in its method, to a later style influenced positively, both in its content and in its method, by the work of Russell and Moore.

A closer understanding of his scheme is gained first from *Some Dogmas of Religion* and secondly from his complete metaphysical exposition, *The Nature of Existence*. Except for *Some Dogmas of Religion*, much of McTaggart's work prior to *The Nature of Existence*, is concerned with Hegel. His first book, *Studies in Hegelian Dialectic*,[14] is an interpretation and defence of the general principles of the dialectical method. *Studies in Hegelian Cosmology*[15] deals with special applications of the method. *Commentary on Hegel's Logic*[16] is a rigorous exposition of the Hegelian process from the category of Pure Being to the Absolute Idea.

Some Dogmas of Religion is McTaggart's atheistic statement of how we exist as spiritual selves in a determined yet self-governing, loving universe. It is the most convenient and accessible synopsis of his thinking. The first chapter, 'The Importance of Dogma', is based on the article 'The Necessity of Dogma', published in January 1895.[17] The third chapter, 'Human Immortality' first appeared in 1903[18] and the fourth chapter, 'Human Pre-existence'[19] in 1904. Both were published together in 1915 as *Human Immortality and Pre-Existence*.[20] In addition there are chapters on 'Free Will', 'God as Omnipotent' and 'Non-omnipotent God'. McTaggart hoped that *Some Dogmas of Religion* would appeal to a wide reading public and become, what Broad eventually called it, a 'semi-popular book'.[21]

Though, like Berkeley, McTaggart thinks matter, though there, is not provable, he believes the essential quality of reality is that of 'perceiving' and not 'being perceived'. Reality therefore exists in the nature of the perceiver and not in the act of perceiving. In this way it somehow lends reality to that which is external to it.

McTaggart argues that science is incapable of establishing dogma. He defines metaphysics as 'the systematic study of the ultimate nature of reality', dogmas as 'all propositions which have any metaphysical significance' and religion as 'an emotion resting on a conviction of a harmony between ourselves and the universe at large'.[22] Religion is not synonymous with morality, it is not simply feeling good or being in awe of the universe nor is it dependent on a god of science or theology, its foundations lie in love for others which itself incorporates both reality and harmony. Even though there is some consensus in human views on morality this is not so with dogma and because of its necessary role in the practice of, and belief in religion, this only leads to increased religious confusion.

The abstract nature of scientific propositions, limited by materialism and mechanics and drawn from only a small universal sample, cannot decide on any absolute truth in the universe and so cannot be religious dogmas. Only the practical application of metaphysics can interpret meaningfully the findings of science and save us from the anxiety caused by its failures. Though a religion based on science rather than dogma would be more empirically certain and subject to greater consensus, pressing questions of meaning that positively affect our search for happiness and harmony still would remain unanswered.

McTaggart holds that metaphysics is the only way to escape the 'dreamy possibilities' of god and because most people do not undertake metaphysical study they have no right to accept religious dogma. However, as religious authority, and the intrinsic acceptance of dogma declines, more people will be unable to hold religious dogma and this will lead to increased loss of truth and a commensurate growth in evil.

Nevertheless, if the universe as a whole is good, it could provide a satisfactory basis for religious belief whether people believed in it or not and ultimately belief would increase, for example if immortality is true we will be immortal whether we believe it now or not. Also, any unhappiness brought about by loss of belief in religion would have compensations, for example, freedom from worship.

McTaggart concludes, after finding common reasons for holding dogmatic belief unacceptable, that dogmas are not subject to the changing nature of our daily lives and, though demanding moral qualities of us and faith enough to trust our conclusions, can only be established by reason without any underlying motives, in other words by the practical application of objective metaphysics. He proposes that metaphysics may become of the greatest practical importance when people realise they cannot have religion without it. Indeed, it may become as certain as physics, immortality being on a par with say the law of gravitation. If so, people would then not have to investigate it themselves as they would have ready-made metaphysical dogmas upon which to rightly base a religion. But, in a world of constantly overturning of philosophical opinions this seems only a distant hope.

Even so, he is cautious about the results of an outbreak of metaphysics for even if the individual tests out metaphysical dogma for himself his results may for a long time still lack the consensus which is so important for knowledge. The route to true dogmas is not at all certain, we may die before we attain them or falsely believe we have attained them and never know it. The need for continuing existence therefore seems unavoidable and the question of immortality remains our primary metaphysical concern.

Geach, in his analysis of McTaggart's system, urges us to take the broadest view of events in history and wisely warns against calls for 'agonizing reappraisal' driven by the irrationality of the 'emotional impact of what is near to us in space and time'.[23] *Some Dogmas of Religion* deflects us from the influences of currency and fad. It explores the nature of our deeply felt confusions and attempts to expose truths that are often masked by our everyday concerns and influences. It causes us wonder if we are becoming disproportionately convinced by the dominance of science and the sprawl of information it presents and whether this will only take us a step nearer despair. It challenges us to consider if we are prepared to run the risk of subordinating the possibility of immortality and the fabric of love to the spiritually unproductive style of modern philosophy. It awakens us to the necessity of pursuing truth by metaphysical argument and the rejection of uninformed and ill-founded belief.

Whereas *Some Dogmas of Religion* tells us that there is nothing in common-sense to make us disbelieve the existence of a mental self, *The Nature of Existence* tells us what self is and fits it into a universal scheme of selves. It is an essentially deductive argument based upon the most meagre self-evident truths: that something exists, that we know that something exists and that what exists is differentiated. By adopting these as a priori necessities McTaggart completely sidesteps epistemology and aims straight for an absolutist metaphysical truth. This is combined with an inductive element: reality must be like this, the only thing I can thing of like this is *x*, so probably reality is *x*. The process leads to the conclusion that the self is real and will continue *sub specie temporis* ('under the appearance of time') as immortal, until the *C* series, which is misperceived as time, reaches the final stage of timeless eternality.

It has been common philosophical practise to look only at Chapter 33 of *The Nature of Existence*, 'Time', or perhaps at the origin of this chapter, McTaggart's article, 'The Unreality of Time'. Since Broad, only Geach,[24] King-Farlow,[25] Shea,[26] Welby[27] and Wisdom[28] have looked further. Understanding that McTaggart's denunciation of the reality of time is only part of a greater conception, is crucial if we are to understand his intention. Farmer thinks that though it is useful to see the temporal paradox in light of McTaggart's fuller philosophical concept, it can be considered quite independently.[29] Farmer's 'limited qualification' to this is that Chapter 33 of *The Nature of Existence* is substantially the same as 'The Unreality of Time', and as it is only in *The Nature of Existence* that McTaggart reflects his more mature philosophy, the chapter on time is only a part of that more mature view. This is wrong. McTaggart's search for ontological meaning led him to the abandonment of time and this search was well advanced by the time he wrote 'The Unreality of Time'.[30] Criticism of McTaggart's work shows how sharply focused on the denial of the reality of time it is. This has caused additional problems. Le Poidevin says of Farmer's scrutiny of commentators' views, 'Many people would be frightened and bewildered by Farmer's McTaggart, and come away with their impression confirmed that time is an intractable field to be approached only by formal logicians and theoretical physicists'.[31] This is the twofold problem that faces us here; not only is time confusing but anything connected with its denial seems to make things worse.

2.2 McTaggart's terms

Before going any further, it is important to be clear about McTaggart's terms. McTaggart calls the series ordered by the relations 'past, present and future' the *A* series and the series ordered by the relations 'earlier and later' the *B* series.[32] The real ordering series that we misperceive as the *B* series he calls the *C* series.[33] The increments of our misperception of the *C* series he calls the *D* series.[34] McTaggart uses the term 'substance' to stand for what is commonly called a 'particular' and the term 'characteristic' to stand for what is commonly called 'universal'. Geach likens McTaggart's argument for substance to Wittgenstein's argument for objects that must be substances of the world.[35] Though he notes that unlike Wittgenstein's objects, McTaggart's substance need not be (and is anything but) simple.[36] On the issue of time and its replacement I think it is satisfactory to refer to substance as anything 'existent'. Broad (who is not sympathetic to McTaggart's views) provides extensive and sometimes elaborate comments on *The Nature of Existence*. He replaces some of McTaggart's terms with terms of his own. In particular he replaces 'perception' when used in the extended sense, with 'prehension', further dividing this into 'w-prehension' and 't-prehension'. 'w-prehensions' always come at one end of the timeless *C* series that is misperceived as a time series and they always appear *sub specie temporis* at the latter end of time. 't-prehensions' are prehensions that are not 'w-prehensions'. Geach thinks these terms relate to sense data and therefore materiality and have no place in McTaggart's system.[37] I retain 'perceive'. For McTaggart 'spirituality' means 'non-material', it does not have any theological connotation. He describes our conscious being as 'self' thus avoiding the problems of 'soul'; we *are* a self whereas we tend to think of ourselves *having* a soul.

2.3 McTaggart's system

In this section I provide an overview of McTaggart whole system as well as a summary of the important features of his fuller philosophy as found in *The Nature of Existence*, without which it is not possible properly to appreciate his atemporal system.

2.3.1 An overview of McTaggart's system

McTaggart's idealistic universe composes the simple substance of selves, the characteristics of which are qualitatively defined in a hierarchy of parts according to their type and degree of perception. Self-perception is primary, perceptions of others are various grades of secondary and perceptions in time are fragmentary. Primary perception, as the only quality of selfhood, is known by introspective awareness. Secondary and lower grade perceptions are known via the fundamental relationship of love. Perceptions of selves accumulate as the atemporal world which they constitute proceeds by serial increments to the final stage; an ultimately timeless yet everlasting monistic absolute — the primary whole. Though the ultimate nature of substance is simple, its appearance to us in all pre-final stages is that of unyielding pluralistic complexity of lower grade secondary or confused fragmentary parts.

The *A* series is the transitory relation past, present and future. It is only in the *A* series that change can occur and, if the passage of time presupposes change, then the *A* series is crucial to its reality. The terms of the *A* series are independent phases because change can only happen if something is once future then present, or present then past. At the same time, in order for the temporal movement that change requires to take place, its phases are interdependent and shared. For example, a future moment *will be* past at a later moment or an earlier future moment will be in the *future past* of a later future moment. The *A* series is therefore self-contradictory.

The *B* series is the permanent relation of successive ordering from earlier to later. It is here we have the confused experience of time even though, by itself, the *B* series is not enough to sustain change. The occurrence of change in apparent time relies on the transitory nature of the *A* series, for whereas events 'move through' the *A* series nothing can 'move through' the *B* series. Although our psychological experience of time is in the *B* series we cannot know where we are in time without the 'now' of the *A* series. Because the *A* series is self-contradictory, whatever the *B* series is, it cannot be temporal.

The *C* series is the one-dimensional series of a self's perceptions made up of the relations 'included in' and 'inclusive of'. All perceptions that are initially 'included in' become permanently 'inclusive of' all subsequent perceptions. The *C* series of any self is always misperceived in the *B* series and will itself only be known by a self when the last term of inclusiveness, the last perception, occurs in the final stage. Because the perception series of each self represents a *C* series there are as many *C* series as there are selves.

Any perception had by any self adds to the total of inclusive terms in its *C* series. Although this produces an 'intensive' difference it does not result in any additional terms (in the same way that the difference between one mental state and another does not create a third mental state). However, even though they do not involve any additional *C*-terms the increments of 'intensive' difference imply some sort of 'extensive' difference in magnitude of misperception. Intensive difference can be seen as differences either in the source of perception or in the resulting perception itself (which, if matter is in doubt, may be differences only in the perception itself) which do not produce a third quantitative difference. For example (McTaggart's), the difference between a bright light and a less bright light does not form a third light of differing brightness.[38] However, as McTaggart says, 'In extensive quantity the difference between a greater and a less quantity of the same sort is a third quantity of the same sort'.[39] These extensive terms form the *D* series which is exclusively a misperception series by which we experience our misperception of the *C* series in the *B* series. Only time in the *B* series will support the misperception of inclusion in the *C* series which therefore itself includes our misperception of the *B* series in the *D* series. Our apparent *B*-experience of the world in all pre-final stages therefore always forms the misperceived increments of the *D* series.

Because it remains a possibility that all selves will perceive all other selves completely in the final stage (although McTaggart feels unable to say that this will or must be the case either directly or indirectly[40]) the *C* series must account for mutual perception. For a self to perceive another self completely it must perceive it clearly and distinctly. This means that even though it is more likely that the system is finite and has a last term, the *C* series of any self must be capable of perceiving all its parts to infinity. Because the *C* series is a simple, one-dimensional series of 'more inclusive' perception, it cannot deal with the possibly infinitely regressive nature of mutual perception inherent in the perception series. A self *G* must be capable of perceiving a self *H* itself perceiving *G* which includes *G*'s perception of *H* and so on. For this to happen the *C* series relies on a system of perpetual determinancy that is capable of infinite internal description — the principle of *Determining Correspondence*.

Determining Correspondence can imply a sufficient description of a possibly infinite substance without presupposing one. If it were presupposed, it would imply a sufficient description without implying what it was, and this would cause a vicious infinite regress. *Determining Correspondence* is a two-dimensional system: one dimension being the terms of the grades of perception and the other the terms of the parts of those grades. In *determining correspondence*, one part is determined by another part and that determining part includes, as part of itself, the determination of the other part. If a sufficient description can be had for a primary part then, by *determining correspondence*, a sufficient description can also be had for all involved secondary parts of any grade. In other words, anything that is itself perceived, has perceptions that have perceptions and so on, and even though these implicit lower grade perceptions are not directly perceived, by *determining correspondence*, they can be included as completely described in higher grade perceptions. *Determining correspondence* precludes error because its parts must, by differentiation of parts to infinity, be distinguishable from one another and must also, by definition, satisfy the requirement to be known completely.

Although *determining correspondence* can describe all parts of any lower grade perception, and there are no such perceptions that cannot be determined by it, the self, as the primary part, even though part of *determining correspondence*, has perceptions in time which are highly confused and unavoidably erroneous. Because these perceptions cannot be known completely they cannot be determined by *determining correspondence*. Parts of perception that are not clear and distinct in this way, and which form the confused differentiation of parts themselves, constitute '*fragmentary parts*'. As the *D* series is the series of increments of misperception of the *C* series in the *B* series, so *fragmentary parts* are the confused differentiations in *determining correspondence* that, because they are in time, are always misperceived. All our experience in all pre-final stages is *sub specie temporis* and therefore always subject to the confusion of *fragmentary parts* and the increments of misperception in the *D* series. To add further uncertainty, throughout all pre-final stages degrees of perception 'oscillate' in clarity.

Assuming the *C* series to be finite, the last term is the only place where there is no misperception. Here, as there is no future and no present, there is no *A* series and therefore no erroneous experience of time in the *B* series. Consequently, as there are no increments of misperception, there is no *D* series nor are there any confused *fragmentary parts*, though the whole itself is a term in *determining correspondence*. At the final stage, the self will not cease to exist — it cannot as there is no later, or more properly 'more inclusive' position for it to occupy — instead, it will appear to exist endlessly and to all intents and purposes will remain 'atemporally immortal'. Such endless existence of the self in an atemporal world is not a contradiction, as when *fragmentary parts* no longer exist the falsehood of time is exposed and 'endless eternality' takes on a different meaning. It is endless because no further perceptions are possible and timeless because all perceptions are completely clear and distinct and preclude the misperceived increments of the *D* series and the *fragmentary parts* of *determining correspondence*.

2.3.2 The self (and the way it is perceived)

According to McTaggart, if atemporality is to be meaningful it must include (or be based upon) a view of our self as a perceiving thing. This departs from our normal view of conscious selfhood which involves the misperception of time.

The *really true* (what I call the true nature of 'time' in the atemporal world. See § 6.6) is made up of what McTaggart describes as selves that are substantial but non-material. (Geach thinks McTaggart's conclusion about immateriality is as paradoxical as his argument about time[41]). It is the perception of these selves one to another, to a greater or lesser degree, that form the *C* series. The *C* series is the series of perceptions that are in the perceiver. The perceptions that *G* has of *H* are perceiving a self, a part of a self, a group of selves or parts of selves. Not only does *G* perceive *H* as being *y* but *G*'s perception of *H* perceives *H* as being *y*. (Geach thinks this latter category of perception should be more properly called 'represent'[42]).

Perception is the only quality of selfhood. It is known to us by introspection and not by our perception of the external world (that is 'external' to perception itself). Some apparent experiences may appear to us to be experiences and may be experiences of the real, whereas some experiences may appear to us to be experiences and are only experiences of the apparent. Our experience of time falls into the latter category. McTaggart thinks that all empirical knowledge is either perception or knowledge based on perception.[43] Such perception is superficial in that it only gives us knowledge about the characteristics of existent substance. It is knowledge by acquaintance with the substance itself and cannot give us knowledge of any real characteristics of the substance.[44] When we perceive characteristics of a substance, we only perceive the particular substance's perception of the characteristics of other particular substances and so on.[45] A self's perception therefore includes the perception of another self that it perceives.

McTaggart acknowledges that it is commonly held that perception cannot be totally erroneous. Either it must be a perception of some sort or what is perceived must actually have the characteristics that we perceive it as having. He believes both these options to be untrue. He thinks that our perception of time, sensa and the nature of the spirit highlight the difference between appearance and reality. This difference is not mistaken judgement but errors of perception.[46] Consequently the existent is other than we perceive it to be.

According to McTaggart something exists; its true nature lying in both its existence and its qualities. Quality is indefinable (but examples might be goodness, happiness, redness or sweetness[47]) but whatever exists will have a plurality of qualities. All qualities themselves will also have qualities to infinity. Substance is related to its qualities and its qualities are related to it. Two substances cannot be completely similar and each substance must therefore be capable of an exclusive description. To describe substance he invokes what he calls the 'principle of dissimilarity of the diverse'. This principle (in some aspects originally formulated by Aquinas) is drawn from Leibniz's 'principle of the identity of indiscernibles'. On McTaggart's principle substances are defined by their characteristics and not identified with them. Such characteristics are sub-divided into qualities and relations. From this he infers the 'principle of sufficient description' upon which every substance must be different from another substance by at least one attribute. It must also have an exclusive description (not simply naming) which is so specific that it is a sufficient description.[48]

Only substances can have the quality of having content and the content of any substance (based on the perception principle) is spiritual. Therefore selves, parts of selves, groups of selves and groups of parts of selves all have content and so are spiritual. Neither matter nor sensa can be shown as having content so are not spiritual and do not exist.[49] Selfhood is the quality of perceiving oneself, therefore, every self-conscious being directly perceives himself in this way.[50]

McTaggart believes he can provide a sufficient description of spirit by showing it to have parts within parts to infinity. For McTaggart, sufficient description in this way is essential to prove reality. (Geach says that McTaggart defines persons by the 'simultaneous fulfilment of ... three principles: sufficient description, ontological determinacy, and infinite description'[51]). McTaggart considers the case for spirit is far stronger than it is for matter or sensa, neither of which can be shown to have parts within parts to infinity and that only pre-existing bias in favour of our everyday, sensory experience would encourage us to think otherwise.

McTaggart's description of the self demands much of us. The self relies upon our acceptance that selves are mental states that form the whole content of selves and that there is no material characteristic associated with mental states. Indeed, we must accept that there is no material characteristic in the universe. The common view these days denies the idea of spirituality, though the basis of this denial is drawn from the apparent connection between mental states and brain processes of which little is understood. Spiritual selfhood is a reasonable alternative to the ambiguities of the supposed material world and though a central state materialist would reject this as superstition, it is not so far removed from the contemporary scientific world-view as we might at first think.

The alternative dualistic view put forward by Geach, that it is more reasonable to accept that mental states somehow combine mental and physical characteristics,[52] is more difficult to defend. After all we continually see the physical world dissolving into the 'spirituality' of science and mathematics.[53] Geach's physicalism leads him to believe that mental states (which implies everything) are entirely reducible to brain states.[54] Geach thinks that there is some sense to McTaggart's assertion that all a self can amount to are the various mental states of a person.[55] He thinks a person may consist entirely in his mental states, but that this is not a reason to conclude that everything that is a self is not material or that sensations of materiality are misperceptions. He does not accept that a self can be wholly immaterial. He feels that McTaggart comes to this conclusion by confusing the terms 'part', 'whole' and 'content'. Geach disputes whether a description of parts of a whole can be the same as the described whole itself.[56] McTaggart's assumption that everything is spiritual is, according to Geach, 'a great weakness in McTaggart's philosophy', and comes from McTaggart's 'assumption that such terms as 'past', 'whole', 'content', etc. are univocal'.[57]

However, I think scepticism opens both alternatives. We can neither justify nor prove spirituality but this does not mean it dissolves into materiality (though of course it might). The reason we 'favour' materiality is perhaps more to do with contemporary scientific influence than reasoned inclination. I think that accepting both that persons may in many ways be mental states while not rejecting materialism, is a view in need of continual adjustment. For example, if the scientific 'goalposts' change, the rules will alter and this view must change. It is therefore always subject to empirically based discovery. I think standing 'outside' these rules offers us more. 'Neutrality' is always relative. McTaggart is not neutral because he does not use known parameters against which to judge his 'neutrality', he is impartial and impartiality does not demand any coordinate system. His arguments are always applicable in a variety of universal systems. For example, it does not matter to McTaggart's system whether the universe is finite or infinite in either direction (though there may be a preference). Any truly meaningful metaphysical investigation must rest on such impartiality.

There is an understandable temptation to liken McTaggart's theory and Leibniz's theory of monads:[58] both dispense with epistemology and make straight for metaphysics, the identity of indiscernibles (dissimilarity of the diverse) is crucial to both and arguably Leibniz too thinks that monads exist outside of time. At the same time the theories are very unalike and, in our context here, in these ways. Each monad perceives all other monads from its own point of view. This view will differ from the view of other monads according to degrees of clearness. Each self consists in its differentiating groups that need not include all selves in the universe. One self may therefore be different from any other self. In addition monads really do change whereas selves in the atemporal *C* series undergo qualitative differences that are misperceived in time.

Selves are primary parts of the universe and their content is what is spirituality. The secondary parts of the system are the perceptions had by selves. Only selves are percipients and only selves can be substantial. They are the centre points of, on the one hand, the world of perception, and on the other, the only value that makes up the absolute entity. Spirituality then *is* the universe and selves are the total content of that spirituality and the only medium by which the spirituality can be known. The self is the simple quality of being capable of perceiving itself. When self-consciousness occurs then the self is plainly evident. This self-consciousness is a substantial awareness that, by its self-perception, proves itself every time the self-perception occurs. Such self-perception will not prove that a self exists, only that something is being perceived as being a self. Such perception could be illusory but McTaggart believes that there is a stronger case for selfhood than there is for sensa or matter.

McTaggart says he owes the reason for believing that the self is known to itself by direct perception to Russell.[59] It was suggested to him in Russell's article 'Knowledge by Acquaintance and Knowledge by Description',[60] though by the time McTaggart wrote this Russell had abandoned the view.

According to McTaggart, asserting the proposition, 'I am aware of equality' (whether true or false), and knowing what it means, means I must know each constituent of it. As 'I' am a constituent of the assertion then I must also know 'I' in some way. 'I' can only be known by description if an exclusive description can be given. Because it is likely that someone else can be aware of equality, then an exclusive description of 'I' in this way is not possible. Nor is it possible to provide an exclusive description of 'I' by introspective awareness of my awareness of equality. Here I cannot be sure that the person making the judgement (that the person aware of the awareness is the same person as the 'I') is the 'I' I am trying to describe. After all, how do I know that the person making the judgement is the same as the person being described? All I know about the person who is making the judgement is the description 'the person making the judgement'. As Kaplan points out, it is difficult to get further than this when, 'Your utterance of the same sentence, or mine on another day, would not express the same content'.[61] Or to extend the sense of this, if no one were to utter this token, 'I would not exist' would be true.

For McTaggart such a description must be an exclusive description, that is, known only to myself. This rules out many of our descriptions as in most cases of description, a similar description is available to others. But though I am aware of describable qualities that can be shared by others, others cannot share my 'awareness' of being aware of these commonly describable qualities. This introspective awareness is attributable only to 'me'.[62]

This introspective awareness does not include 'awareness of being aware'. If I am aware of *X* and aware of being aware of *X*, I am judging that the 'I' who judges that 'I' am aware of *X*, is the same 'I' who judges that I am being aware of someone, who may be the same 'I' as I think I am when judging, being aware of *X*. This is known by description. If, on the other hand, I know myself by acquaintance, by directly perceiving myself, then I am justified in believing that the 'I' who is aware, is the same 'I' who judges the 'I' as being aware of myself being aware. Such introspective awareness of our own mental states justifies our belief in our own existence.

Only selves, as Primary Parts can be percipients. All the content of the universe is made up only of Primary Parts. The universe is therefore made up, in its entirety, of selves. As no self forms part of another self neither can God be a self nor can the universe (unless the universe is the content of one mind). Any self is only part of the whole but the whole cannot be a self. This applies whether or not there are any other selves in the universe. As all spirit falls within the content of selves, then selves form a set of parts that are the total of all existent spirit. As spirit is the quality of all content then selves together form the totality of all existence.

Selves can be sufficiently described according to the degree or intensity of love they hold for another self.[63] A differentiation by quality could provide an infinite number of degrees of intensity if there were an infinite number of selves. If the number of selves were finite there could be at least a sufficient number of degrees to provide a qualitatively different description for each of them. He thinks that love is never 'in respect of' qualities of the beloved but is often 'because of' the qualities of the beloved.[64] That the essence of love 'springs from a sense of union with another self'.[65] In other words, selves are the fundamental things and love between them is the fundamental relation.[66] This is difficult to understand. It is hard to see how love could isolate individuals because love is a universal relation; the spirit that loves *x* (where *x* is some spirit that is loved) could never provide a sufficient description because it names another individual.

McTaggart believes it likely that there is a tonal variation of qualities that may be infinite. This provides a differentiating relationship of the quality of love between two selves that is absolutely exclusive.[67] Though again, this is hard to go along with because loving with tone or intensity could always be experienced by some other spirit. For McTaggart, love is not a constant quality that touches all selves; it is a vital and 'created' exclusivity between selves. Because of the possible minute variations in lovingness between selves, they can experience a love that is theirs and theirs alone. No other love will be the same as a love that exists between two selves. Love in general will always be the same qualitative experience, but the constituents of the experience of love will always be different. As the *C* series proceeds towards the end-stage, the degree of perception between selves increases as does the fullness of love.[68]

Value will increase as the number of incidents of value increases. Because time is misperceived the repetition of values such as pleasure or pain will not become wearisome. It will only add to the total sum of the value of pleasure or pain.[69] Value is represented by quality and not quantity or duration.[70] Value is not the possession or maintenance of high levels of good. It is the increase in the amount of good or a reduction in the amount of evil. The conclusion of the *C* series will bring the union of the total of all value in the series at that stage, plus the total increase in value throughout the series. In the final stage, the values of love, knowledge and virtue, will be at their highest because the *C* series will be in a state of the greatest unmixed non-hedonic good. The apparent evil we see in the present is our misperception of reality. This, in turn, leads us to mistakenly believe in what we see as present and past. The argument for ultimate good though, has no firmer foundation than the idea of heading towards an immortality of evil. The idea of heading towards an immortality of good is a 'pleasant' one, whereas the idea of heading towards an immortality of evil, is an 'unpleasant' one.

2.3.3 The general nature of spiritual existence

For McTaggart, existence is a quality that belongs to particular things and events. The whole nature of the existent is a simple quality that contains a complex of other qualities. Although characteristics may appear as complex, the whole nature is one of simplicity. The existent may grow in complexity, as a multitude of perceptions each plays a part in its construction, but its ultimate complexity becomes a quality that is simple. The quality of the whole nature, although including all the characteristics and events that make it up, defines the existent in terms that include, without further reference, all the parts. We may assess the quality of the person without paying attention to the quality of all their various parts. In the same way, we may assess the simple quality of the whole, without concerning ourselves with the associated subordinate qualities, characteristics or events. These qualities stand in relations to each other. The entities (which themselves possess qualities and stand in relations to other entities and to the whole), are the particulars, themselves neither qualities nor relations, which McTaggart calls substance.

McTaggart finds it impossible to define exactly the qualities of existence or relations between such qualities. But he thinks it is possible to define the quality of being a substance and any particular can be accorded this definition. The substantial reality is therefore made up of all particulars that are any states, events or things. Any class, member of that class or group to which those members or classes belong, are substances. Each substance is independently different, has a different quality and stands in a different relation to any other substance. No substances can be the same. They must have qualitative differences based upon the peculiar relations they hold to all other substances. This means that diversity of substance necessitates qualitative difference. In other words, the diverse are dissimilar.

McTaggart calls the substance itself a Primary Whole. A substance that exists in relation to the Primary Whole is a Primary Part of the Primary Whole. When what would otherwise be a Primary Part, corresponds to a Primary Whole in its relationship to it, so as to effect its description, it is called a Secondary Part of the First Grade. Where there is what would otherwise be a Secondary Part of the First Grade, which affects the description of a Secondary Part of the First Grade, it is called a Secondary Part of the Second Grade. If *A* is a Primary Whole and *B* a Primary Part, *B!C* stands for that part of *B* which stands for *C* and therefore stands for a Secondary Part of the First Grade, and *B!C!D* would stand for a Secondary Part of the Second Grade (see § 2.3.4 for a full explanation of the symbol '!').

Any change in any part of any characteristic in the universe means that none of the substances that exist prior to the change will exist after the change. The whole substantial universe will be extrinsically altered by the slightest change in the nature of any Secondary Part of the Lowest possible Grade. In order to understand the nature of any one substance it needs to be described. An exclusive description, which allows us to identify that particular substance as different from any other particular substance, need not be a complete description of the substance itself. Though, as more substances need to be described, the complexity of the description grows. In addition, much of the description will be in terms of the relational qualities of other substances. For the self to be real it must be capable of complete description. A complete description requires that a substance have parts within parts to infinity. At first sight it appears that there is no possibility of such a full description without invoking a vicious regress. To provide this full description McTaggart comes up with a principle of perpetual determinacy he calls the 'principle of determining correspondence'.[71]

2.3.4 The principle of determining correspondence

The principle of *Determining correspondence* is an attempt to provide a complete description of parts within parts to infinity.

McTaggart offers five different forms of perception or belief about perception, all of which involve us in misperception.[72] 1. The perceptions of existents as matter or sensa (which they are not). 2. The belief that no perception perceives selves as perceiving other selves and their parts (though they do). 3. The belief that we can have perceptions of judgement, assumptions, imaginings, awareness of characteristics (all of which are themselves perceptions). 4. The belief that volitions and emotions are perceived as judgements, assumptions and imaginings (though they too are perceptions). 5. The perceptions of existents in time (though there is no time). As far as '5' is concerned, there is no time so any temporal perceptions need to be reinterpreted as part of the inclusive *C* series. As for '1' to '4', though these appear to be misperceptions they cannot be because the self can only have true perceptions. By *determining correspondence G!H!K* is a complete perception of *H!K* (the complete description of *H!K* includes all parts of *H!K* such as *H!K!M!P* and so on) and therefore cannot misperceive in any of the above ways 1 to 4.[73] The remedy for '1' to '4' therefore rely upon our acceptance of the system of *determining correspondence*.

Determining correspondence allows that without me being aware of it in its exactness, something that I perceive is in fact being perceived exactly. In other words, as a self I will have no content except my perception of myself and another self. My perception of another self must be absolutely clear and distinct because this perception must be capable fully of reflecting all parts of all my perceptions. On the face of it, sufficient description is only possible if, and only if, the descriptive series is finite. For example, if *A* depends upon *B* for its description and if *B* follows *A*, then a sufficient description is only possible once the nature of *B* is known. Clearly if the series is infinite there can never be a sufficient description of any of the preceding terms.

The dilemma that leads to the system of *determining correspondence* is similar in form to the dilemma about time. It is obvious that something exists but our usual attempts to describe it fail, therefore the way in which we habitually describe it is erroneous. This error can only be avoided by taking another view. This other view does not seem to accord with our common sense view, but it does escape the original problem.

A repeating one-to-one correspondence such as '*M* is *N*', 'it is true that *M* is *N*' or 'it is true that it is true that *M* is *N*', cannot describe the series of parts of *A*. Neither can using the relating terms of *M* and *N* having relationships to the new terms of '*M* being equal to *N*', for not only has *M* a relationship both to *M* and its relationship of equality to *N*, but also *N* has a relationship to *N* and its relationship of equality to *M* and so on. This second alternative fails because there is no uniform relationship between the terms of such a series and the parts of *A*. This is because the fresh terms are not part of the terms that preceded them. They are, as McTaggart says, 'intercalated' between all the terms of the previously generated stages and cannot therefore determine the series of the parts of A.[74] Nor can the infinite series of sets of parts be found in propositions or characteristics that are not simple. Propositions have limited scope for analysis and characteristics are only made up of simple parts that cannot be analysed any further.

Determining correspondence works like this. 1. Assume that *A* has a set of parts *B* and *C* (this could be a set of parts to infinity). Now assume that *B* and *C* (and so on) has a set of parts corresponding to each set of parts of *A*. 2. Assume that the correspondence is a one-to-one relation between members of the sets of parts. A sufficient description of *C* includes the fact of its relation to part of *B* and therefore determines a sufficient description of that part of *B*. 3. When one part is determined by another part, that determining part includes as part of itself the determination of the other parts. This can be expressed as *B!C* (a secondary part of the first grade) for that part of *B* (a primary part) which corresponds to *C*, and *B!C!D* (a secondary part of the second grade) for that part of *B* that corresponds to that part of *C* that corresponds to *D*. On this, if a sufficient description can be had for primary parts then a sufficient description can also be had for all secondary parts of any grade. In other words, anything that is itself perceived, has perceptions that have perceptions and so on. Therefore any perception has as part of it all these other perceptions, although they are not directly perceived.

A substance *C*, which is in some relation to part of a substance *B* (if perceived), is in a relation of *determining correspondence* if *C* intrinsically determines a sufficient description of *B!C* by determining a sufficient description of all the parts of *B!C* to infinity.[75] Each perception in the universe increases the number of *determining correspondence* relations and ultimately all selves will stand in *determining correspondence* relations to all other selves. This does not imply a vicious circle. In a universe of just two selves, *A* and *B*, the relations would consist in *A*'s perception of *A* and *A*'s perception of *B* together with *B*'s perception of *B* and *B*'s perception of *A*. Such perceptions would be entirely clear and distinct and would reveal as further parts, *B*'s perception of *A*'s perception of *A*, *B*'s perception of *A*'s perception of *A*'s perception of *A* and so on. Such 'inner differentiation' has no finite limit.

Determining correspondence provides for a sufficient description of one set of parts that implies a sufficient description of all lower grade secondary parts to infinity without presupposing them. If we did presuppose them it would imply that there was a sufficient description without implying what it was. It would therefore cause a vicious infinite regress. It would be impossible for it to have what McTaggart calls a 'total ultimate presupposition' while not being able to be without one.

One problem with *determining correspondence* is that it seems to disallow what Geach calls 'discursive thought'.[76] Because everything in the system of *determining correspondence* is necessarily a direct perception there is no provision for consequent thoughts about the perception itself. Broad[77] says that McTaggart embarks upon the principle of *determining correspondence* because he holds that there is misperception in our perceiving (in Broad's terms 'prehending') certain particulars as selves and our perceiving certain particulars as experience of perceptual cognition.[78] Geach points out that Broad's comment (Geach calls the comment 'absurd') that McTaggart's proof does not account for repetition of terms, is meaningless. If repetition occurred, it would simply lead to a vicious circle as opposed to a vicious regress.[79] He says, 'McTaggart, like Wittgenstein in the *Tractatus*, is committed to the view that the Universe could be *completely* described in completely general terms, with no irreducible use of a 'non-connotive' proper name; and I think this view ought to be accepted'.[80]

Any determinable characteristics that are perceived must be perceived in a certain determinable form (for example, if I perceive colour then I perceive the form 'redness'). Broad holds that McTaggart may have forgotten that there can be, and are, misperceptions because I may perceive an object and then misperceive its determinate form. For example, I may perceive the colour to have the value of redness, but if misperception is possible then I may misperceive its value as 'blueness'. According to Broad, only if the argument against the misperception of determinable characteristics is set aside is it possible to accept that the grading of perceptions (from first grade secondary parts perceiving primary parts of selves and second grade secondary parts perceiving first grade secondary parts as selves), is valid. Broad thinks that in order for a first grade secondary part to be perceived as a self, and as part of a whole being perceived via *determining correspondence*, McTaggart has to show that the perception must not be a group of perceptions or part of a perception. This, he feels, McTaggart does not do, nor even attempt in the case of perceiving only as a part of perception. He concludes that McTaggart has not provided any proof that a higher grade part perceives a lower grade part as a 'w-prehension'. Stebbing,[81] feels that although Broad shows McTaggart's fuller endeavour to be unsuccessful, he does not emphasize enough what she considers McTaggart's desire to, 'use logic to establish his mystical conclusions'.[82] Wisdom (using Broad's argument mainly) finds McTaggart mistaken in holding that the only indicator of the necessary hierarchy of sufficient description is a primary quality or a quality dependent upon a primary quality.[83]

Determining correspondence is not the most convincing of McTaggart's argument. However, we may (without being apologetic) consider what it is he is trying to achieve. He is trying to link the reality of the *C* series and the everyday misperceptions that we experience in the *B* series. In the *C*-world I may have a direct self-perception when I reflect on some matter or another. However, in the face of everyday experience this 'discursive thought' seems unavailable by direct perception. Perception is measurable against the workings of our own minds. On balance, I think it is more difficult to accept a principle that conflicts with these workings, than it is to accept a principle that conflicts with something like our everyday view of time.

Determining correspondence is important to the *C* series more in respect of our ability for infinite perception within an infinite series, than to support the reality of the series as an inclusion series with a finite limit. Its reduced importance in the latter context means that although we need to understand the workings of the principle (as it is always present in McTaggart's system), the *C* series does not collapse without it. It may fail us if we wish fully to work through McTaggart's system (and have to account for the possibility of an infinite *C* series), but it does not fail us if we use it only in the sense of a perceptual descriptor of a finite series.

2.4 Conclusion

In this chapter I have assumed that it was necessary to introduce McTaggart. I have described his work, some of his terms and some important features of his fuller philosophy. I believe this overview shows that his denunciation of time, though it can be properly judged as an argument in its own right, has only limited meaning if not placed in the context of his fuller metaphysical exposition. McTaggart's argument against time is not an isolated argument but part of a fuller vision of reality. Only with this appreciation can we evaluate the atemporal world he describes and so fully understand the reasoning behind, and the consequence of, his assertion that time is unreal.

Notes

1 J.McT.E. McTaggart, *The Nature of Existence* (Cambridge University Press, 2 vols., vol. 1, 1921; vol. 2, ed. C.D. Broad, 1927). All references to *The Nature of Existence* take into account amendments recommended by Broad (C.D. Broad, *Examination of McTaggart's Philosophy* (Cambridge University Press, 2 vols., vol. 1, 1933; vol. 2 (Parts 1 and 2), 1938), vol. 1, p. lvi, vol. 2, Part I, p. lxxxv), and Keeling (S.V. Keeling, Notes, 'McTaggart's *[The] Nature of Existence*, Vol. 1, Comments and Amendments', *Mind*, vol. 47 (1938), no. 188, pp. 547-50).

2 J.McT.E. McTaggart, *Some Dogmas of Religion*, 2nd edn., ed. and with Introduction by C.D. Broad, including McTaggart's list of emendations extracted by S.V. Keeling, new Introduction by G. Rochelle (Bristol, Thoemmes Press, 1997. First published London, Edward Arnold, 1906; 2nd edn., 1930).

3 J.McT.E. McTaggart, *Philosophical Studies*, ed. and Introduction by S.V. Keeling, Foreword by Gerald Rochelle (Bristol, Thoemmes Press, 1996. First published London, Edward Arnold, 1934).

4 J.McT.E. McTaggart, 'Mysticism', *The New Quarterly*, vol. 2 (1909), no. 7, pp. 315-39. Reproduced in *Philosophical Studies*, pp. 46-68.

5 J.McT.E. McTaggart, 'The Unreality of Time', *Mind*, vol. 17 (1908), no. 68, pp. 457-54. Reproduced in *Philosophical Studies*, pp. 110-31.

6 J.McT.E. McTaggart, 'The Relation of Time and Eternity', an address before the Philosophical Union of the University of California, 23 August, 1907, *University of California Chronicle*, vol. 10 (1908), no. 2, pp. 127-52. Reprinted as a pamphlet (Berkeley, Calif., The University Press, 1908).

Reprinted in *Mind*, vol. 18 (1909), no. 71, pp. 343-62. Reproduced in *Philosophical Studies*, pp. 132-55.

7 J.McT.E. McTaggart, 'The Meaning of Causality', *Mind*, vol. 24 (1915), no. 95, pp. 326-44. Reproduced in *Philosophical Studies*, pp. 156-78.

8 J.McT.E. McTaggart, 'Personality', *Encyclopaedia of Religion and Ethics*, vol. 9, ed. J. Hastings (Edinburgh, T. and T. Clark, 1917), pp. 773-81. Reproduced in *Philosophical Studies*, pp. 69-96.

9 J.McT.E. McTaggart, 'The Further Determination of the Absolute'. Privately published 1893 (reproduced in *Philosophical Studies*, pp. 210-72); 'An Ontological Idealism', *Contemporary British Philosophy*, 1st Series, ed. J.H. Miurhead (London, George Allen and Unwin, 1924) pp. 251-69 (reproduced in *Philosophical Studies*, pp. 273-92); 'The Individualism of Value', *International Journal of Ethics*, vol. 18 (1908), no. 4, pp. 433-45 (reproduced in *Philosophical Studies*, pp. 97-109).

10 J.McT.E. McTaggart, 'Propositions Applicable to Themselves', *Mind*, vol. 32 (1923), no. 128, pp. 462-4. Reproduced in *Philosophical Studies*, pp. 179-82.

11 J.McT.E. McTaggart, *Dare to be Wise* (London, Watts and Co. (for The Rationalist Press Association), 1910), 9pp. Reproduced in *Philosophical Studies*, pp. 37-45.

12 J.McT.E. McTaggart, 'Introduction to the Study of Philosophy' (McTaggart's personal notes for the course of Friday evening lectures held at Trinity College, Cambridge, from 1899-1914). Reproduced in *Philosophical Studies*, pp. 183-209.

13 This is evidenced in McTaggart's drafts of *The Nature of Existence*, originally titled *The Dialectic of Existence*. The existing copies of Draft *A* containing Books 5, 6 and 7, are titled *The Dialectic of Existence*, as are Chapters 35 and 36 of Draft B. The remainder of Draft *B* and the chapters of Draft *C* which are available are entitled *The Nature of Existence*. Chapters 35, 'Sense Data' and 36, 'Spirit', are derived from the Chapter entitled 'Spirit' in Draft A. In Draft *C* and in the published work Chapter 35 eventually becomes, 'Sensa'.

14 J.McT.E. McTaggart, *Studies in the Hegelian Dialectic* (Cambridge University Press, 1896).

15 J.McT.E. McTaggart, *Studies in Hegelian Cosmology* (Cambridge University Press, 1901).

16 J.McT.E. McTaggart, *Commentary on Hegel's Logic* (Cambridge University Press, 1910).

17 J.McT.E. McTaggart, 'The Necessity of Dogma', *International Journal of Ethics*, vol. 5 (1895), no. 2, pp. 147-62.

18 J.McT.E. McTaggart, 'Some Considerations Relating to Human Immortality', *International Journal of Ethics*, vol. 13 (1903), no. 2, pp. 152-71.

19 J.McT.E. McTaggart, 'Human Pre-existence', *International Journal of Ethics*, vol. 15 (1904), no. 1, pp. 83-95.

20 J.McT.E. McTaggart, *Human Immortality and Pre-Existence* (London, Edward Arnold, 1915).

21 C.D. Broad, 'The Local Historical Background of Contemporary Cambridge Philosophy', in C.A. Mace, ed. *British Philosophy in the Mid-Century* (London, George Allen and Unwin, 1957), p. 43.

22 McTaggart, *Some Dogmas of Religion*, §§ 1-3.

23 P.T. Geach, *Truth, Love and Immortality, An Introduction to McTaggart's Philosophy* (London, Hutchinson, 1979). All references to Geach are in accordance with his own errata (letter, P.T. Geach to G. Rochelle, 6 September 1990).

24 Geach, *Truth, Love and Immortality*, p. 176.

25 J. King-Farlow, 'The Positive McTaggart on Time', *Philosophy*, vol. 49 (1974), no. 188, pp. 169-78.

26 W.R. Shea, 'McTaggart and the Neo-positivist Entropists', *Philosophy*, vol. 50 (1975), no. 193, pp. 346-51.

27 V. Welby, 'Mr McTaggart on the "Unreality of Time"', *Mind*, vol. 18 (1909), no. 70, pp. 326-8.

28 J. Wisdom, "McTaggart's Determining Correspondence of Substance: A Refutation', *Mind*, vol. 37 (1928), no. 148, pp. 414-38.

29 D.J. Farmer, *Being in Time: The Nature of Time in Light of McTaggart's Paradox* (Lanham, Maryland, University Press of America, 1990), p. 7.

30 See Rochelle, *The Life and Philosophy of J.McT.E. McTaggart, 1866-1925.*

31 R. Le Poidevin, Book Review, 'D.J. Farmer, *Being in Time: The Nature of Time in Light of McTaggart's Paradox*', *Mind*, vol. 100 (1991), no. 3, pp. 388-90.

32 McTaggart, *The Nature of Existence*, § 306.

33 Ibid., § 347.

34 Ibid., §§ 571-72.

35 L. Wittgenstein, *Tractatus Logico-Philosophicus* (London, Routledge and Kegan Paul, 1986. First published 1922), §§ 2.0201-0202; 4.2211.

36 Geach, *Truth, Love and Immortality*, p. 44.

37 Ibid., p. 139.

38 McTaggart, *The Nature of Existence*, § 377.

39 Ibid.

40 Ibid., § 439.
41 Geach, *Truth, Love and Immortality*, p. 74.
42 Ibid., p. 124.
43 McTaggart, *The Nature of Existence*, § 299.
44 Ibid., § 301.
45 Ibid.
46 Ibid., §§ 302, 691.
47 Ibid., § 60.
48 Ibid., § 104.
49 Ibid., § 381.
50 Ibid., § 382.
51 Geach, *Truth, Love and Immortality*, p. 130.
52 Ibid., p. 133.
53 See R. Penrose, *The Emperor's New Mind* (Oxford University Press, 1989) and H. Putnam, *Reason, Truth and History* (Cambridge University Press, 1981).
54 Geach, *Truth, Love and Immortality*, p. 133.
55 Ibid.
56 Ibid.
57 Ibid., p. 134.
58 See G.W. Leibniz, *The Monadology*, trans. by R. Latta (Oxford, Clarendon Press, 1925).
59 McTaggart, *The Nature of Existence*, § 382.
60 B. Russell, *Mysticism and Logic* (London, George Allen and Unwin, 1918), p. 211.
61 D. Kaplan, 'Thoughts on Demonstratives', in P. Yourgrau, ed. *Demonstratives* (Oxford University Press, 1990), p. 36.
62 McTaggart, *The Nature of Existence*, § 383.
63 McTaggart discusses his view of the nature of love and its relation to the perceptions of selves in *The Nature of Existence*, §§ 455-81.
64 Ibid., § 465.
65 Ibid., § 464.
66 See W. J. Mander, 'On McTaggart on Love', *History of Philosophy Quarterly*, vol. 13 (1996), no. 1, pp. 133-47.
67 McTaggart, *The Nature of Existence*, § 486.
68 Ibid., § 473.
69 Ibid., § 868.
70 Ibid., § 869.

71 Ibid., §§ 197-8.
72 Ibid., § 505.
73 Ibid., § 547.
74 Ibid., § 196.
75 Ibid., § 202.
76 Geach, *Truth Love and Immortality*, p. 131.
77 Broad examines the principle of *determining correspondence* in his *Examination*, vol. 1, pp. 373-400 and criticises it in vol. 2, pp. 203-21.
78 Broad says this 'is analogous to the legal maximum that an accused person is to be deemed innocent unless he can be shown to be guilty'. (Ibid., vol. 2, p. 203). This may be the statement that prompted Geach to compare McTaggart to the Tichbourne claimant who, because he was an honest man, requires a fair hearing. (Geach, *Truth, Love and Immortality*, p. 8).
79 Geach, *Truth, Love and Immortality*, pp. 52-4.
80 Ibid., p. 54.
81 L.S. Stebbing, Critical Notice, 'C.D. Broad, *Examination of McTaggart's Philosophy, vol. II*', *Mind*, vol. 48 (1939), no. 190, pp. 214-20.
82 Ibid., p. 220.
83 Wisdom, 'McTaggart's Determining Correspondence of Substance: A Refutation'.

3 McTaggart's Denial of the Reality of Time

3.1 Introduction

In this chapter I describe McTaggart's argument against time. In doing this it will be seen that *B*-theory answers to McTaggart's claim that time is unreal, leave the *B*-theorist denying time as we experience it, in much the same way as McTaggart. This means that although *B*-theory arguments against McTaggart's position may have strength, they do little to move forward from a position of temporal scepticism.

In § 3.2 I explain McTaggart's argument against the reality of time. In §§ 3.2.1, 3.2.2 and 3.2.3, I describe the three steps that lead McTaggart to conclude that for him the crucial feature of time, the *A* series, is contradictory. I then go on to show that this presents us with psychological problems. In § 3.2.4 I show how we must come to terms with the idea that our experience of change in the *A* series is wrong. In § 3.2.5 I show that, on this view, our experience of time in the *B* series is also wrong. In addition I show that trying to uphold the *B* series as an alternate basis for the reality of time still is denying time as we know it. The nature of change lies at the root of McTaggart's denial. In § 3.3 I show that there are some reasons, even in the face of most opinion to the contrary, to accept that it is possible that it is events that happen to things that change and not things themselves that change. In § 3.4 I show that the argument for the vicious regress of the *A* series is not so easily answered as some think. To do this, in §§ 3.4.1, 3.4.2, 3.4.3 and 3.4.4, I analyse four arguments commonly used to attack it. In § 3.4.5 I show that atemporality is a reasonable view to hold, although common sense tells us otherwise. In § 3.5 I conclude that though common sense makes us suspicious of the unreality of time, we find enough reason (in McTaggart and our inability to convincingly answer him) to be strongly suspicious of its reality. On this basis alone we have good reason to investigate McTaggart's idealist replacement for time as we know it.

3.2 McTaggart's argument against time

In this section I describe McTaggart's argument against the reality of time. I describe the three steps that lead him to conclude that for him the crucial feature of time, the *A* series, is contradictory (§§ 3.2.1, 3.2.2, 3.2.3). I then go on to show that this presents us with psychological problems. In § 3.2.4 I show how we must come to terms with the idea that our experience of change in the *A* series is wrong. In § 3.2.5 I show that, on this view, our experience of time in the *B* series is also wrong. In addition I show that trying to uphold the *B* series as an alternate basis for the reality of time still is denying time as we know it.

By defining time in a certain way and using McTaggart's argument, we can conclude that time is unreal. This is not startling as time is a common source of confusion although McTaggart's method and conclusion seem paradoxical as an argument against time conflicts with something apparently incontrovertible. Philosophers have devoted much energy trying to show McTaggart's denial (in some degree or another) to be meaningless. This usually takes the form of denying the *A* series and upholding the view that, in the *B* series alone, a de-tensed world is satisfactory. However, this position still is denying time as we experience it and puts the *B*-theorist in the same position as McTaggart. Others think the argument a logical manipulation; Gotshalk thinks that McTaggart makes the assumption that whatever appears contradictory or impossible to thought must be either non-existent or unreal.[1] This is unjust, McTaggart's claim against the reality of time is neither glib nor intentionally paradoxical. It is McTaggart's firmly held belief that the real world cannot be represented in time as we experience it.

McTaggart believes that it is possible to prove that nothing existent can possess the characteristics of being in time. He asserts that time is unreal and therefore all statements that involve time are erroneous. He acknowledges that this is paradoxical and departs from our natural position.[2] Indeed, it departs further than does the idea of the unreality of space or matter. Whereas our own introspective states are not material or spatial, they are unavoidably temporal. Nevertheless, as McTaggart says, 'in all ages and in all parts of the world the belief in the unreality of time has shown itself to be singularly persistent'.[3] He thinks that, 'Such a concurrence of opinion is highly significant, and is not the less significant because the doctrine takes such different forms and is supported by such different arguments'.[4] As well as the philosophy of the East and West, religion and mysticism, he cites particularly the opinions of Spinoza, Hegel, Kant and Bradley, but says that his reasons for believing that time is unreal are different from any of them.[5]

McTaggart's argument against time has two parts. Part one involves three steps. Step 1. Time involves change. Step 2. Change can only occur if there is a series of relations past, present and future, the *A* series. Step 3. The *A* series involves a contradiction that leads to a vicious infinite regress therefore there can be no time. Part two argues that the relation earlier and later, the *B* series, must be temporal, but that there can be no *B* series without an *A* series. As the *A* series is self-contradictory there can be no *B* series. This leads to two problems. Problem 1. Our view of things changing in the *A* series is wrong. Problem 2. Our experience of time in the *B* series is also wrong. These are problems because nothing seems more certain than our experience that things are changing in time.

The 'tensed' view considers past, present, and future as prior to earlier and later, so that all meanings of earlier and later are explainable in terms of past, present, and future. The 'tenseless' view on the other hand, considers earlier and later as prior to past, present, and future so that all meanings of past, present, and future are explainable in terms of earlier and later. The 'tensed' view is called the '*A*-theory' and the 'tenseless' view the '*B*-theory'. In everyday life there is little gained from *A* series and *B* series distinctions beyond distinguishing events or moments as having two very different types of feature; *A*-features for events or moments that have a past, present, future relation and *B*-features for events or moments that have an earlier, later relation. Though there are times when it is useful to refer to the *A* series as 'tensed' and the *B* series as 'tenseless'. In addition, as Geach points out, the ambiguity of the ordering relation of the *A* series and the fact that the term 'series' can be either singular or plural, are of no real concern.[6]

3.2.1 Part 1. Step 1. Time involves change

It is generally accepted that time involves change of some sort and that there could not be time without change. Aristotle says, 'Now everything that changes changes in time',[7] and 'not only must that which is changing have changed, but that which has changed must also previously have been changing, since everything that has changed from something to something has changed in a period of time'.[8] There seems no doubt that change is, in some way, going on.[9]

3.2.2 Part 1. Step 2. Change can only occur in the A series

In our everyday lives we think that we perceive all events in time as present. In fact this is far from the case and it is our 'experience' of present in the *A* series which provides both our sense of change and a feeling of confusion about a place in time. Broad thinks that any phase of time can be perceived ('prehended' as he terms it) as a temporal whole. That degree of presentness perceived will tail off, from a maximum at its later boundary, to a minimum at its earlier boundary. Because of this tailing off, the degree of presentedness will accordingly diminish. Importantly, perception of any temporal whole only ever takes place in the past. The perception phase is completely past when perception of it first begins. Broad thinks that the notion of temporal transition derives from our experience, in this way, of the specious present. The tailing off of presentedness from earlier to later, the continually overlapping specious presents and the reduction in perception of any one past, conspire to form our apparent experience of 'now'. This means that between any two specious presents there will be an infinite number of other specious presents.[10] We need not accept the notion of the infinite divisibility of specious presents to accept the common sense of the rest.

If we do not perceive an event as present then we think we remember it as past, infer it as past or infer it as future. This set of mental relations always forms an *A* series.[11] In our everyday lives we do not experience time as a *B* series. We think only of past, present and future and only use the *B* series relation to establish which events happen before or after other events. We may say some event was earlier than another event, or some event was later than another event, but such thoughts always rely upon the more fundamental *A* series. *A*-features seem inescapable in a temporal world. McTaggart thinks that the distinctions of positions in time in the *A* series may be an illusion and that the real nature of time lies in the *B* series. If this were so, we should not perceive time as it really is, though we might be able to think of it as it really is.[12]

We identify closely with the present because it is the only state that we directly perceive. Because of this, we think of our present state as moving towards later events in the *B* series and as moving towards the future in the *A* series. In effect what is happening is the *A* series is moving along the *B* series from earlier to later and the *B* series is moving along the *A* series from future to past. It is like Zeno's rows of soldiers filing past one another in opposite directions, though not with the same spatially paradoxical consequences.

McTaggart has a *B* series view but he does not accept that the *B* series is fundamental to a time series. He thinks the *A* series is the fundamental series. So for any event that is earlier or later than any other event to be comprehensible, it will always need to be placed in an *A*-relation. To establish temporal ordering, it is not enough to know that a clap of thunder is associated with a flash of lightning. Such a relation need not be the 'later than' relation: it may be another form of causal relatedness. For temporal ordering, what we need to know is whether that particular flash of lightning, has a past, present or future relation to its consequent clap of thunder; the event itself cannot enable temporal ordering. However, the fact that it is an event that I am considering, means that it automatically has a temporal sense. Nothing can be an event unless it occupies a place in the *A* series; though we shall see later that *B*-relations are much closer to the reality of what is apparently going on in time than *A*-relations.

The *A* series forms our understanding of what we think is happening in time. The *B* series, which is a series relative to it's own terms, is of no use to anyone in that world. To make sense of the progress of events, the observer has to be able to fix on some known temporal position. It may be my present experience as my specious present or it could be in some way present without being my specious present. It may be an event that has a known position in the past of the *A* series or may even be something in the future of the *A* series. Though here, to be of any use, we would need to know with certainty how long it would take us to arrive at this future point and doing this would mean such a future reference point was based on a point in the present.

A *B*-world is not sufficient to provide us with an adequate understanding of time as we experience it. If we try to construct such a world we soon become involved in *A*-terms. For example, if event *M* is earlier than event *N*, we are on fairly safe ground. *M* will always be earlier than *N* and *N* will be always later than *M*. If we want to locate *M*, all we need to do is to locate *N*, then look for an event that happened earlier. If there are only two events, *M* and *N*, *M* is found straight away in its 'earlier than *N*' position. In this relative *B*-world we could always find the other term providing we knew the whereabouts of one. It would be the same no matter how many events there were. All we would need to know would be which events any of them were earlier than or later than. This would, of course, be a very complicated temporal system to compute. However, in the simple 'two event-world' if we lost track of both *M* and *N* (I am not sure what we would know if we did lose track of the only two events in the world) then we would never find either of them again. Even if we lost track of say *M* and knew that *N* still existed, we would never locate *N* because *M* is no longer available as a reference point. If there were more than two events, say three, then this third event, *O*, would need somewhere to (relatively) be. If *O* is earlier than *N* and later than *M* we can also always find *O* providing we know where either *M* or *N* is. However, if we lose both *M* and *N* we will never be able to locate *O*. It is clear that in a *B*-world of a greater number of events we would need to know the relative whereabouts of a very high number of events to establish the position of only one. We would be faced with the disastrous consequence of being unable to completely calculate our place in our own history.

No matter how many events are in the system, they are only ever in respect of other events. This does not seem enough to constitute time. Time is more than just relativity of events. It has to have a known point of reference that can always be found and is always independent of any relative relations in the *B* series. We could take some particular event, say the 'Big Bang' as a reference point but we would not necessarily have escaped the problem. If the 'Big Bang' is the first moment in time we would be on quite safe ground, but if it is not, then all we have done is to select a relative event in the *B* series and use it as a convenient means of forming a pseudo-independent reference point in the *B* series. If we are unsure whether it really was the first moment in time then using it has no fundamental sense; it is only an arbitrarily chosen point of convenience that saves the *B* series from endless self-referencing.

Our specious present usually fills the role of reference point. However, the *B* series cannot admit our specious present as a reference point for two reasons. First, if all the other events are later than our specious present, still we would need to place these events in earlier and later relations one to another to know how far in the past they are. In doing this, we invoke an *A* series. Secondly, the specious present itself is part of the *A* series. It does not form part of an earlier or later relation.

Why do we assume so strongly a reality of time in the *A* series when we also suspect that it is illusory? We assume the reality of the *A* series because, if we wish to live in a temporal world, we cannot do without it. However, though it seems indispensable, McTaggart thinks that it is, in truth, illusory. Our conscious perception of movement in time is thus at odds with any objective reality we incorrectly (though excusably) credit it with. If there is to be any reality to time, it must consist only in the *B* series, that is, in some tenseless form.

What are the consequences of a *B* series without an *A* series? We have accepted, as a basic premiss, that change is necessary for time and there is no reason to abandon this. Therefore, if the *B* series is to be shown as temporal, change must be evident in a *B* series without an *A* series, in just the same way as it is to be found in the *A* series. But the *B* series seems incapable of sustaining change. Any event that has occurred in the *B* series will always have occurred. It will always have the same position in time. It will always have been earlier than following events and later than preceding events. Events in the *B* series always have a permanent position. This is so whether they merge into succeeding events or are moments in absolute time. Only if the characteristics of an event could change could there truly be said to be any change within the permanent *B* series cannot change. On the face of it (though later we shall see that this is a superficial and incorrect assumption) an event is its own defining characteristic. But our mistake is to extend its permanent placement in the *B* series to a description of a permanency in the nature of the event itself. If event *a* is earlier than event *b*, then its only temporal characteristic is that it is earlier than event *b* and this will never change. But its temporal characteristic — its place in the *B* series — does not define the nature of the event itself. We cannot say that an event *a* is also an event *b* but we can say that event *a* later becomes part of a larger event *b* and when this happens the characteristics of event *a* itself have changed even though its *B*-position has not. Because such a change has not happened in the *B* series it must have happened in the *A* series.

Welby, for example, thinks that change can only be in the characteristics of events as they are presented in the *A* series.[13] This is partially correct and rightly transfers the place of change to the transient *A* series but it is incorrect in assuming that it is the placement in time that actually changes the characteristics whereas it is the characteristics themselves that change and this is enabled because of its placement in the *A* series. Characteristic *a* at t^1 does not itself change into characteristic *b* at t^2 (as though it were a thing that was altering its state) but it is *a* at t^1 and *a (changed)* at t^2 and *a (changed again)* at t^3. This has happened because t^1 was once in the future then in the present and then in the past and this is only in respect of something against which the *A*-process can be measured. However, t^1 will always be earlier than t^2 that will always be earlier than t^3 and so on.

There is only one way in which the characteristics of an event can be said to change and that is in respect of the *A* series. Change will only occur in an event that was once future, then present and then past. In the future, it changes as it moves from the distant future to the near future. In the past, it changes as it moves from the near past to the far past. If there is no real *A* series within which this change can occur then there cannot be said to be real change. As long as we accept the need for change there seems no getting away from it; the *B* series is not sufficient to form time because it will not sustain change.[14]

McTaggart thinks that in reality, the *B* series cannot be anything other than temporal. Its connecting terms are time related. But as it cannot, by itself, sustain change, it must depend upon an *A* series to be real. Evidently there can be no time without an *A* series and there can be no *B* series without an *A* series. To prove that time is real we would need to prove that the *B* series is sufficient to sustain change. As the *B* series alone cannot sustain change, we must therefore discount it as a means by which we can prove time. Our only hope for a real meaning to time is in the *A* series.

In the *A* series past, present, and future are all characteristics ascribable to events or to moments in time. They may be relations (which is most likely) or qualities (which is less likely) but if they are relations they will generate relational qualities in each of their terms. It does not matter whether they are relations or qualities because in either case still they would be incompatible characteristics.[15] Anything that is in the *A* series must have a relation to something outside the time-series.[16] Relations that are exclusively within the time-series can never change and therefore must form part of the *B* series. The *A* series is therefore not only transitory but relative. It is not meaningful to conceive of 'past' unless it is 'past relative to something subsequent'. This relation will vary because something that is now 'past' was once future in relation to something that is external to that series. For example, 'ten o'clock tomorrow morning' is in my future now and the day after tomorrow it will be in my past. The demonstrative 'I' therefore provide the external relation. The context-sensitivity, or indexicality of this type of external fixed point of reference, is essential to meaning in any relational time series.

Gotshalk considers relations in the *A* series are permanent.[17] Gale reduces the *B* series to an impure *A* series, 'which is determined not only by an unqualified past, present, and future but also by more past and more future'.[18] Grünbaum thinks this non-metrical concept of overlapping temporal intervals can be described as 'more past' or 'more future'.[19] McTaggart accepts the concept of 'more future' but only when he has firmly denounced time in favour of the *C* series and can describe its parts in relation to the 'whole' which lies at what is the equivalent in the *C* series to the future. I think relations in the *B* series are unequivocally permanent, this applies to moments in their relation to one another (if moments are separate realities) and of the relation of events to moments. Unlike *A*-features, *B*-feature relations are always explainable in terms of themselves, for example, tomorrow will always be later than today. As McTaggart says, 'Two events are exactly in the same places in the time-series, relatively to one another, a million years before they take place, while each of them is taking place, and when they are a million years in the past'.[20] We do not have to accept the detailed placement of events in a fixed 'later' to accept the inevitability of events in a fixed 'earlier'.

If the *B* series is permanent then the changing relations of events of the *A* series, must be relations to something outside the time series. *A*-features do not exist independently, pastness has no meaning unless it is in relation to presentness or futurity.[21] By choosing to separate time into *A*-features and *B*-

features and accepting that change is fundamental to time, we have to accept that time can only realistically be found in the *A* series. I do not pursue the idea of a separate reference point for the *A* series, though it is one sometimes argued for in defence of the *A* series being non-contradictory. McTaggart ignores this. Any external reference point can never be rid of an inherent regression to another sub-*A* series and so on and any argument in its defence only seems to duplicate the original problem.

3.2.3 Part 1. Step 3. The A series is contradictory

The *A* series has as its terms the three independent phases past, present and future and these terms are interdependent. However, each term in it is incompatible with the other two terms. An event is only always either past, present, or future. It cannot be more than one at any one time. This exclusive nature of the terms is essential to change. Each event is associated with it at least two of either past, present and future for change to take place. If the time-series has a first moment, that moment (of change) will only be present and past. If the series has a last moment, that moment (of change) will only be future and present. But still this is inconsistent with the obvious incompatibility of the terms.[22] Nothing could be more different than past, present and future are to each other.

If we accept that the *A* series is essential for time to be real, we must also accept that in the most obvious way it 'does not work'. McTaggart's argument has led us in a circle. To find out why this has happened we need to look more closely at the nature of this incompatibility. We use tensed language to describe events as either past, present, or future. If an event is either past or present or future there is no problem because each term is successive. Our language has no form to describe the three terms as simultaneous. There is no reason it should; after all, no event is ever past, present, and future simultaneously. What we need to know is what is the temporal meaning of 'has been', 'will be', and 'is'? The terms past, present, and future may be indefinable, but the nature of the terms still involves a contradiction. If we attempt to remove this contradiction McTaggart thinks we become involved in using the same terms and this leads to a vicious infinite regress. This does not arise from the impossibility of defining the terms, but from the nature of the terms themselves. They involve a contradiction and any attempt to remove the contradiction involves the use of the same terms and so on.[23]

When we say that an event *X* 'has been' *Y*, we are saying that the event was *Y* at a moment in the past. When an event *X* 'will be' *Y*, we are saying that the event will be *Y* at a moment in the future. When we say that an event *X* 'is' *Y*, then we are saying that the event *Y* is in the present.[24] Some event that 'is' present, 'will be' past and 'has been' future, means that it is present at a moment of present time, past at a moment of future time and future at a moment of past time. This applies equally to moments or events; if they are present they were once future and will, at some future time, be past.

If an event is present then there is no moment of past time at which it is past. But there are moments of future time in which it will be past and these moments in the future, which will be past, are equally, moments of past time. Present moments cannot be past in a past time, they can only be more distantly past. Similarly, a moment of future time will eventually be a moment in the present and then a moment in the past. As a moment moves from the far future to the near future to the present, it takes a place in the 'future present' and the 'future past'. It is therefore present and past at different moments in future time, though it cannot be present or past at different moments in past time.[25]

From this a further contradiction arises. Any moment that is in the future is also a moment in the 'future present' and the 'future past'. No moment strictly has a place in the future unless it is also present and past. Every moment that we say, at the present, will be future in a future time, will also be present then past at a 'later', future time. This difficulty persists infinitely and is, according to McTaggart, a vicious infinite regress.[26] It can be argued that this is relying too strongly upon the 'river of time' principle, that it is placing too much certainty on a future being available from which will 'flow' more future presents. However, if we think of these future presents as in the very near future — the very next moment so to speak — they are readily acceptable temporal features. Of course, if the universe ended in the next few moments, there would not be much of a supply of future moments (and our optimism would have been misplaced), even so, still there may be time for us to appreciate that some immediately past moments, were very recently, immediately future moments. Though indexically now in the past, this impression still would suggest that this 'passage' occurs.

Any apparently separate phase of change in the *A* series cannot be described without recourse to the other phases. The defined *A* series requires these separate phases of change to be independent phases. But based on this definition apparently they cannot be separate and can only have meaning in relation to each other.

If the *A* series collapses because of internal contradiction then the idea of change, and consequently time, looks very shaky. The validity of the *B* series relies upon the reality of time. If time is not real it is difficult to afford any reality to the *B* series. If this is so, then nothing can be said to be past, present or future, earlier or later, or simultaneous with anything else. Consequently, nothing really changes. If this is the case then it is unlikely that *A*-characteristics can occur, and if they do not then time must be unreal. According to McTaggart this is not simply a paradoxical conclusion but proof of fact which forces us to accept that our experience of time must be wrong. When we perceive things in time we are, as he says, 'perceiving it more or less as it really is not'.[27] If, like Dummett[28] and Geach,[29] we can support this we must look for reality in an atemporal universe that is, in some way, 'neutral' to these implications.

3.2.4 Part 2. Problem 1. Our experience of change in the A series is wrong

Facts are facts about events. If events do not change then facts do not change. According to McTaggart real change occurs only if events change and any change in any event, no matter how seemingly insignificant or unrelated, changes all other events in the universe (and any facts about them).

Atemporality requires a special kind of change because our normal view of time — that it is somehow 'produced' by change — is redundant. If time is unreal then the change, which is undeniably happening, must be different from how we usually think of it. This alternate view of change is now generally called *Cambridge change*, a term invented by Geach because he considers it common property of Cambridge philosophers of McTaggart's era.[30] McTaggart says, 'if anything changes, then all other things change with it. For its change must change some of their relations to it, and so their relational qualities. The fall of a sand-castle on the English coast changes the nature of the Great Pyramid'.[31]

As we have seen, we experience time as it presents itself in the *B* series and this is reliant upon the *A* series. But if the *B* series is not real, what it presents to us as temporal must also be unreal; what it is that we think of as time must be something else. However, this is meaningless if we accept conventional change because it only produces or supports conventional time. If, with McTaggart, we accept the redundancy of the *A* series our conventional view of change is no longer tenable. We must therefore try to find the real quality of what appears to us as actual change. Common sense might suggest that the presence of actual change is stronger evidence of the reality of time than a negative argument about the terms of time is evidence of its unreality. To proceed with McTaggart we must set aside such phenomenological common sense.

Cambridge change enables us to go beyond the confusions provided by the *B* series. To fit the system *Cambridge change* must be possible without the *A* series but on the face of it, no change can take place in the *B* series alone. Any event that was earlier than any other event will always be earlier than that event because the *B* series is a permanent series. We could try to get round this by imagining that a certain event 'becomes' another event but this does not work. If *M* becomes *N* then there would have been a moment when *M* ceases to be an event and *N* starts to be an event. As such a ceasing-to-be and starting-to-be does not constitute becoming events *M* and *N* remain distinct events. To blur this distinction is only to introduce a paradoxical 'heap' of time which is at some time both *M* and *N*. If we introduce an absolute time external to the events in question we would get no further. In fact things might be worse for as well as the events still being placed in a permanent relation one to another they are also in an equally permanent relation to absolute time.

The death of Queen Anne (McTaggart's example) can only change in relation to the terms of the *A* series. The causes of the death, its effects, that it is a death, that it is the death of Queen Anne, all remains permanently fixed in the *B* series. The only way it changes, is that once it was in the future, then it was in the present and then it was in the past. The only ways that the characteristics of an event can change, are when they are determined by the terms of the *A* series. If there is no real *A* series then there can be no real change. Because there cannot be time without the *A* series and the *B* series alone cannot uphold time and consequently change, the nature of change (as we conventionally determine it) must also be wrong. Any propositions in the *A* series can only be said to change if they are sometimes true and sometimes false. This is so of propositions that deal with the place of anything in the *A* series. For example, 'the Battle of Waterloo is in the past' and 'it is now raining' are both in this category. Outside this category, propositions do not apply to change. Actual change is constituted of two propositions about an object. These propositions differ in that one refers to an earlier time and one a later time and that one is true and the other is false. For example, a statement such as 'the grass is green' at time t is true and 'the grass is green' at time t^{l} is false, it having become browned by frost (Geach's example). Such a comparison refers to an actual change in the object. If on the other hand we said, 'the grass is greener than the hedge' at t and this was true and 'the grass is greener than the hedge' at t^{l} and this was false, because of the browning by frost, we have a different conclusion and a different view of change. Although the grass still has undergone an actual change, on McTaggart's principle, so has the hedge. Whereas once it was less green than the grass, now it is more green than the grass (the grass having become browned by frost). Not only has actual change taken place but so has *Cambridge change*.

Cambridge change reveals a change in an object because it has an altered relation to anything else. This is irrespective of any actual change that has or has not occurred in the object itself. On this principle, even after my death, if anyone thinks something about me, 'I' change, because after this thought, 'I' have become 'me' that has as part of it that new thought. In fact, after my death I will continue to change as everything else in the universe changes because every change will alter the relation that those elements of the universe have to 'me'; once I was 'my life plus *x* events after my life', then I was 'my life plus *xy* events after my life' and so on. What is true of me, in apparent totality, in the penultimate moment of my life, is only true of me until the next moment that occurs after my death. I will continue to change quite irrespective of whether I am alive. This meaning of change will apply to everything including God, who will continue to undergo change as each self's relation to God alters with every new thought. *Cambridge change* includes any actual change — the change of common sense — which refers to change in the object itself.

According to McTaggart, the reason for our misunderstanding of change lies in the way we perceive actual change. The internal characteristics of an object that exemplify actual temporal change cause us to misperceive change and to misinterpret its true nature. Actual change may seem more tangible than abstract *Cambridge change* but tangibility has no automatic claim to a resemblance to reality. Geach thinks that any confusions about the reality of *Cambridge change* are misunderstandings.[32] Actual change is simply a convenient view of change. Objects that change represent to us the changing nature of the world as we experience it and our experience of actual change forms a sound basis for our daily lives. But such convenience does not represent the truth. The true nature of change is very different from its apparent nature. Though we find our everyday view of change convenient, it is only part of the change that is really happening.

Change can only occur relative to something. It is easy to infer this relativity in the object itself, for example, the grass was once green and now it is less green. In truth the grass, and its greenness, is relative to more objects than just itself in time. Indeed, if it can be judged relatively and if there is no time for it to be relative to, then it must be relative to something else. In *Cambridge change* it is ultimately relative to everything in the universe, in the same way that everything in the universe is relative to the grass. Actual change is only a minor relativistic constituent in *Cambridge change*. That it happens to be the most obvious and easily understood indicator of change and that it is confined to the object that we deem to be changing is not sufficient to define change in itself.

Cambridge change is neutral with respect to the distinctions of *A* and *B*-theory. The *A*-theorist thinks that temporal becoming accounts for changes in time whereas the *B*-theorist thinks that the temporal relation of succession accounts for such changes. But changes would or would not be real, quite irrespective of which view is taken. In either case the meaning of *Cambridge change* in the *B* series would be unaltered. The *A*-theorist thinks that *B*-relations depend on *A*-determinations. The *B*-theorist explains changes in time entirely in terms of *B*-relations. But neither view affects any idea that we may have about there being a definite order of events in time. Neither the *A* nor the *B*-theorist is saying that any event *M*, which was once earlier than an event *N*, will at sometime be later than that event. If *B*-ordering is retained there is no difficulty in extending our view of change. The matter of dependence or otherwise of the *B* series is, as Farmer points out, peripheral to the question of change.[33] Any degree of diversity in how we judge change can be accommodated by *Cambridge change*. Keeling tells us, change 'happens', it is ultimate and therefore indefinable.[34] A view, much influenced by Bergson (the 'modern Heraclitus' as Whitrow calls him[35]) who believes time, and ultimate reality, constitutes the process of continual change.[36] Keeling is no sympathiser of Russell's views and Bergson is the subject of scathing criticism from Russell but all their views can be drawn into McTaggart's idea of a multi-relativistic framework of change. The shortcomings of the Russellian view are exposed only by its lack of dynamic.

We can only describe what it is that happens, we cannot describe the 'happening' of it. The difference of relationship is not a species of change. If things are differently related at different times they should not be said to have changed their relations, but to be said only to be differently related. Any argument against the denial of the *A* series should not rely upon a defence of actual change: it is too limited. Our McTaggartian view of change gives us more scope to look at time clearly. *Cambridge change* is a consequential difference in relationship, consequent that is upon another change that has undeniably 'happened'. Geach correctly calls this view 'the only clear, sharp, conception of change we have'.[37] Any actual change must always be subject to *Cambridge change* and as such is absorbed by it. *Cambridge change* is the only true understanding of change that we can rightly hold.

3.2.5 Part 2. Problem 2. Our experience of time in the B series is wrong

To pursue this I use Russell; not only was he McTaggart's contemporary but much of his thinking still forms the foundation of modern *B*-theory views.[38] The Russellian view holds that consciousness is essential to the existence of the *A* series. An assertion that something is present, means that it is simultaneous with that assertion. An assertion that it is past, means that it is earlier than that assertion. An assertion that is future, means that it is later than that assertion. This means that anything is only past, present, or future, in relation to some such assertion. Without consciousness, which provides the assertion, there would only be earlier and later. Oakeley thinks that if our 'experience' is removed then the *A* series has no meaning. She, like Russell, says past, present and future rely upon consciousness, whereas earlier and later do not. However, she goes further, believing that change is represented by altered states in a permanent relationship of earlier and later. Though she supports McTaggart by recognising that such a series, although appearing as a time series, is in fact not a time series,[39] she holds that there is no change because there is no change in a past that is not 'fed' by present or future, and that the *B* series alone cannot sustain change. On the face of it this is a strong argument. It forces us to recognise the influence of our psychological view and how easy it is to label the universe with our own conscious judgements. It makes us suspect that McTaggart's analysis of time is an expression of the way that we mentally construct the *A* series, or more precisely, the way that McTaggart has constructed it for us.

The idea of there being contradictions in events being both, for example, future and present in future time, are reliant upon our experience in a 'moving' psychological time. If we are not able to think of what the future may hold, or what it may be at all, then such contradictions are irrelevant. However, we must remember, it is only in the *A* series that any change is possible. Without consciousness, there can only be a permanent *B* series relation of earlier and later and consequently no change.[40] Change is irrelevant in an *A* series that has only one 'phase' of change. There is little point in asking the whereabouts of change if an event *N* is in the past and there is only past. *N*'s relation to some assertion *V* would always be that of being past relative to *V*. But the *B* series relation remains. If *N* is in the past relative to *V*, then *N* must have been earlier than *V*. Although *A*-terms do not hold up independently, *B*-terms are inescapable.

It may be too simplistic to consign change exclusively to the *A* series. After all, the *B*-theorist must believe that there can be an unconscious *B*-world of permanent relations that can somehow be deemed to include change. Russell defines change as 'the difference, in respect of truth or falsehood, between a proposition concerning an entity and the time *T*, and a proposition concerning the same entity and the time *T*′, provided that these propositions differ only by the fact that *T* occurs in the one where *T*′ occurs in the other'.[41] But in defining change in this way he involves time in the definition itself. We have the 'entity', which could be consciousness or any relative reference point, but also we have *A*-features and, as long as we consider *A*-features essential for time, we therefore have time. Russell might be rejecting the *A* series, but his definition of change involves it. If it involves the *A* series, it must involve time. If it involves time it must involve change and if the *A* series is rejected, then change must go with it.[42]

Russell's argument does not concern a change in the events in time, but a change in the entity to which the events happen. In other words, the 'thing' or 'object' which is subjected to change. Let us be clear about 'things', 'facts' and 'events'. Although I believe there really is a distinction to be made between a change in the inherent qualities of a 'thing' and an 'event' relating to that 'thing', I do not think there is a meaningful distinction between a 'fact' and an 'event'. They are both ways of describing the state of some thing in a way which is separate from the thing itself. But is it things or events that change?

The Russellian *B*-theorist argument runs as follows. The event of a poker being hot on a particular Monday will always remain an event in the *B* series. It is a permanent *B*-feature. If, prior to this particular Monday it was cold, and since this particular Monday it was cold again, these events would in no way alter the state of the poker itself on this particular Monday, when it was hot. The state of the poker itself will undoubtedly be different at different times but that particular event, on that particular Monday, will never change.

The argument is inviting us to accept that the poker itself may change and yet events related to the poker do not? In doing this it not only asks us to believe something improbable it mixes up the status of 'facts'. It ties together facts about the poker to the thing which is the poker and separates events from them both. Facts are not part of the inherent quality of a thing; they are ways of helping describe events. As if this were not enough it confuses the nature of events and facts with their placement in the *B* series and this is at the root of much confusion in *B*-theory. Events and facts about the state of the poker are all unchanging *B*-features; their placement does not change. Still we are left with the fact that the state of the poker may change. But why should we believe that just because facts or events about the poker are permanently in the *B* series this is more important than that the poker itself somehow changes. The notion of the permanent fixity of events in the *B* series only shows that the *B* series is permanent. If there was nothing more to the universe than this then all earlier events would be identical to all later events; the *B*-theorist argument does nothing to answer the obvious charge that this is not so. The *B*-theory cannot account for changes in things (though it assumes them) or what happens to things to change them and the fact that all earlier events are not identical with all later events exposes the need for an explanation of what is happening that the *B*-theory ignores. The only way that events and facts about the state of the poker can change is if they are in the *A* series; only if the facts can be placed in either the past, present or future can there be any real change. But as we have already seen, events and facts as unchanging *B*-descriptors are not compatible in the *A* series.[43] If, against Russell, we assume that it is events which change and not things but that they are unable to change in time then they must be able to change in some atemporal sense.

The *B*-theorist holds that such statements as 'The Battle of Waterloo is in the past', or 'It is now raining', are ambiguous. Using propositions that are always true or always false, such as 'The Battle of Waterloo is earlier than this judgement' or 'The fall of rain is simultaneous with this judgement', make them definite. If this is so then all judgements are always true or always false and as no facts change there still is no change.[44] But this does little to escape beyond simple *B*-permanency and proposes a world far removed from the world we experience.

We would find it strange to live in such a *B*-world. I would find it difficult to organise my life, saying that every event is in some *B*-relation to some judgement that I make about it. Though granted, if I could cope with this sort of long-winded explanatory process it would enable me to position everything in the *B* series and so know where everything was. But the problem of no change remains. Each fact, because it is either true or false, cannot be, for example, fact *M* that changes into fact *N*. Fact *M* was earlier than fact *N* and both facts *M* and *N* were earlier than my judgement about them. In this fact-centred world, nothing would ever change and yet we know that things do change. The *B*-world uses things instead of events to explain away what change is. On McTaggart's theory, what really changes is the nature of events themselves; facts only help us to describe events and the changes that occur to them.

Nor can I realistically have any notion of the future in such a *B*-world. I can only deal with facts that are earlier than any judgements that I make about them. I cannot consider any potential facts, which are at this moment later than any judgement that I may make about them, which will ultimately affect their truth or falsity. For example, I may know that *M* is earlier than *N*, in respect of the judgements that I make about them. What I cannot know, is any consequence that one fact plus another fact may have. This would result in my having no motivation for action because even though I had memories of previous actions I would have no future position in which I could anticipate their pattern being followed. It is difficult to imagine how we could exist in such a world. It would seem impossible in a *B*-world to get from sitting in front of my breakfast (for whatever reason) to actually eating it. Granted, if I lifted my spoon I could say I had lifted my spoon but I would have no concept, even though I may have done it before, of a future time in which I could bring it crashing down on my egg I cannot know, or suppose, what might happen later than any judgements that I had made because they would be in the future and therefore part of an *A* series. I cannot anticipate events later than the point at which I can judge them in the *B*-world. Although this eliminates the *A* series, which is its intention, it simultaneously eliminates what is a sensible view about our experience of time, which is that events succeed each other in the *A* series. The *B*-world relies on consciousness both for its appreciation and manipulation. It cannot exist in itself as a means of temporal explanation. It cannot support change nor can it explain it away. The *B* series is falsely erected and cannot have meaning as an explanation for time without its partnership with the change-tolerant *A* series. We shall see later that McTaggart's atemporal world performs much better; it is an objective reality reliant upon perception that is not consciousness and includes change as well as our misperception of time.

Upholding the *B*-world model puts the *A* series is in doubt and this has led to an unfounded concentration on problematic *A*-features. The idea being that if the *A* series can be found lacking then by default the *B* series takes on the role of sustaining time. However, we need to bear in mind that the *B*-theorist is usually more concerned with debunking the *A* series and finding temporal reality in the *B* series than finding that time is not real. But what, we may wonder, is the point? *A*-features dominate our understanding of time. *B*-features are not sufficient to keep us adequately informed about time. The *A* series is constructed, by McTaggart, because when it is knocked down the *B* series collapses as insufficient to support time. Disproving the *A* series does nothing positively to support the *B* series and such claims invariably leave us in the same dead end as a *B* series bereft on an *A* series. Arguments which maintain that the *B* series alone is sufficient for time are universally unsuccessful in providing a clear picture of time as we experience it and the unavoidability of a changing world. The only positive step is to recognise the inability of the *B* series to sustain time and instead investigate what time really is. The *B*-theory argument is reactionary to the *A* series contradiction. It is a negative argument which tries to prove temporality based on the *B* series and one that totally fails.

We can now turn to look more closely at two crucial factors. The first involves a further analysis of Part 2. Problem 2., that our experience of time in the *B* series is wrong, in particular, that it can lead us to believe time can happen in a world of unchanging *B*-positioned events while ignoring the demands of obvious change (apparently consigned to things in some indescribable way) which is really brought about by what happens to the nature of events. If successfully defended this will reinforce my assertion so far that because the *B* series alone is untenable, the denial of time, based on the contradictory nature of the indispensable *A* series, is true. The second is to do with Part 1. Step 1. of the argument, in which the *A* series involves a contradiction that leads to a vicious regress. This, if defended successfully, will further uphold the basis of the denial of the reality of time.

3.3 It is events that happen to things that change — things themselves do not change

The nature of change lies at the root of McTaggart's denial. In this section I show that there are good reasons, even in the face of most opinion to the contrary, to accept that it is events that happen to things that change and not things themselves that change. I accept that the placement of events in the *B* series does not change but because this means that the *B* series cannot sustain change it also means it cannot support time. What does change is the nature of events themselves and this can be explained whereas the unfounded idea that things somehow change, even though *B*-facts about them do not, is indefensible. This argument demands that we hold a strong idealist view.

Cambridge change may be acceptable as a general interpretation of change but if it is applied to atemporality it must show that change takes place in events as opposed to things. This is because atemporality relies on a universe of spirit which excludes the objective existence of things and their consequent changing. Any argument that shows that things change would strengthen the physicalist *B*-theory view of temporality. It is therefore important for the atemporal argument to reveal a strong case that it is events that change.

Although we have seen that in *Cambridge change*, if one thing changes all other things change, there is in fact no objective change in those other things. In talking like this we are using 'thing' to stand for any 'thing' to which actual change can be applied; a thing-in-itself or some inherent quality it may possess. If the change is not in the thing, then it must be in something else and I propose that it is in the nature of the event related to that thing or to do with the event of change itself.

The argument that things change has many followers, Russell, as we have seen, leads the argument for things changing, with Broad,[45] Stebbing,[46] Cleugh,[47] Marhenke,[48] Pears,[49] and Smart[50] in hot pursuit, followed by most interested philosophers to date. Geach thinks that it is things that change and that the idea of events changing is introducing a false notion of a 'second-order change' in the *A* series.[51] Few speak up against this, Mink being the most notable.[52]

Prior's argument typifies the *B* Series view on this.[53] It goes like this. Things happen and events happen to occur, events are changes and essentially a process is an event that goes on for a length of time. In other words, changes change.[54] A change in the *A* series is occurring only while it is happening, so a change can only happen to an event while it is occurring. Julius Caesar cannot be getting tired now unless the doctrine of immortality is true. McTaggart's view is that because the doctrine of immortality *is* true, then Julius Caesar is indeed getting tired as each further moment occurs. This is not simply a quirk of his idealist philosophy. It helps us recognise how events relating to spiritual entities convey the meaning of change. *Cambridge change* means that the status of events changes as each new moment is added. It has nothing to do with a person still being alive. But the *B*-theorist cannot accept that the death of Queen Anne (or Julius Caesar) has an indefinably long history in the past, as the death of Queen Anne is a very short bit of history in what, in the *A* series, is a very long history.[55]

B-theory says that when we think we talk about events we are really talking about things. When we think we are referring to changes in events we are really talking about complicated changes in things. This applies to changes in things that are presently going on, just the same as to changes in things that have not yet happened or are no longer going on. However, in a way, this leads us back to *Cambridge change*. If the death of Queen Anne is the subject of our thinking about the death of Queen Anne then the death of Queen Anne has changed. Obviously the death has happened and Queen Anne is no longer alive, so she cannot have more things 'happen' to her but the event of her death has been altered by our thinking about it. But the *B*-theorist would maintain that the changes in Queen Anne that we now talk about, are either not changes 'in the proper sense' or do not fit into the formula of the *A* series.[56] This raises two questions: what is a change in the proper sense and why does a *B*-theory need the support of the *A* series?

The *B*-theorist maintains that if the flow of time implied by the *A* series were true *Cambridge change* might be acceptable, but it would never be accompanied by, what Prior calls, 'genuine change'. On this view *Cambridge change*, as a form of 'general change', is too general to be meaningful as far as the individual facts about Queen Anne are concerned. Though accepting the *A* series it places reality somehow in the domain of the living, experiencing being. So even though it is not Queen Anne's conscious appreciation of what is happening to her that is in question after her death, her death defines her ceasing to be a 'real' thing. When Queen Anne is alive, 'real' things happen to her; when she is not alive only 'quasi' or 'metaphorical' things happen and these cannot be happening to her, or more precisely, cannot be attributed to her. This notion lines up with the idea of temporal becoming. That is that the metaphor of the river of time is inappropriate and time is only a product of conscious experience. In other words, time cannot be imagined unless there is someone to imagine it. It follows that if conscious beings are removed from the *B*-system then no real changes can be happening. This anthropocentric view can be found in Bergson's distinction between the scientific spatialization of time as *le temps* and the conscious experience of time in *la durée.*[57] The idea that time derives from 'my relation with things' is found in Merleau-Ponty.[58] The anthropocentric view and its relation to the *A* series can be found in Grünbaum[59] and Smart.[60] The contradiction is that though such views cannot provide us with any temporal system external to consciousness, they involve 'real' events happening to things which are not necessarily conscious.

At first sight the argument for things changing has obvious appeal. After all we seem to witness the continuing change of every 'thing' that we are aware of, including our selves. The *B*-theorist accepts this common sense ground as the fact, saying that changes do not occur as changes in events and events do not themselves change because events *are* changes. The argument goes like this. If my poker was hot on a particular Monday and is now no longer hot, I am clearly talking about an event that has changed. Even if I did not accept that there had been a change in the poker itself, I must accept that the event of the poker in the past is not the event of the poker now. From this I can infer that change has taken place. The event of the poker in the past is now no longer available to me, whereas the event of the poker now is available to me (which also at this point suggests that the event has changed). The *B*-theorist (realising that he might be supporting the changing of events) then says that I must ask what is the nature of the event? When I consider the previous state of the poker I must decide whether it was the poker, as a thing, which was once hot, or whether it was an event of hotness that could somehow be attributed to the poker. The *B*-theorist thinks there is no contest and comes down firmly for the former. It was the thing itself, the poker, which was hot. It could not be the event that has changed because the event could only be attributed to the thing at that particular time. The event has no usefulness except describing the nature of the thing at that particular time. Once we leave the past moment we 'leave' the event that we were using to describe the thing at that time. Now we are in the later moment, still we are dealing with the 'same' thing, but it can be described by a different event. The thing has therefore changed. It can be described by using a different event, hot or cold, but the event has no identity that can be said to have changed.

This may seem common sense. It seems to consolidate on the obvious and any argument against it appears, at best, ill-advised. But perhaps we are dealing with a case of philosophical weight more than good sense. I think there are sound reasons why we should object to this view.

Objection 1. Objecting to permanency of the distinctions of the *B* series, inasmuch as they appear fixed within a frame of reference, as opposed to remaining permanent without alteration throughout time, is irrelevant to the conclusion that time is unreal. Even if the *B*-features are permanent throughout time, still things cannot be said to change. The poker will always be the poker in the *B* series irrespective of its state. The idea of the poker being a 'changed' poker, just because it is hot or cold means either that it has changed into a different poker or it is the same poker associated with changed events. If it is a different poker then it has not changed and if it is the same poker associated with different events then it is the events that are associated with it that have changed. We may say, 'he is a changed man since the accident', but we do not mean that he is actually a different man. He is the same man but the events associated with him (or facts about him) are different. If he were in fact a different man, then the question of change would never come into it. If the fifty year old man with a beard was suddenly replaced by a twenty year old man without a beard and without any other similar features, we would not be considering how this one man had 'changed'. We would be considering how we could connect the two different men.

Objection 2. There is no sense in re-introducing change into the terms of the *B* series once it has been removed from the relations. This makes change impossible to analyse. *B*-features are changeless because they are unquestionably permanent. To try to force change into the series is an error. Quite simply, if some thing is earlier than another thing it will never be anything other than earlier. Whether it is 'actual' change or *Cambridge change* does not matter of course, because *Cambridge change* includes actual change. It is what it is that we are applying changes to that matters. For example, I may think that things cannot themselves change and that only the events that happen to things can change. But either can only be a possibility in the *A* series. Neither of these possibilities could occur in the *B* series alone because change (no matter what sort or to what it is attributable) cannot occur within a set of relations that are permanent. If change is what tells us that time (or something like time) is happening we are only informed of this by its transitory nature. Change *is* transitoriness and the *A* series is the aspect of time that defines it. The *B* series is the aspect of time that defines permanency from which we adduce history and order. Changes in events can only occur in respect of the *A* series. If an event is fixed permanently in the *B* series its place and state will always remain the same.

Though it may be true that in some circumstances events themselves *are* changes this does not provide a reason for substituting permanent 'things' for changing 'events'. Because this fits with the empirical proposal that things (as molecules) change their state does not mean that it is a reflection of reality. Even if it was 'things' that were changing still they would form a series of changing things and the problem of analysis in the *B* series would remain. The poker is a thing that was once hot on a particular Monday and is now the same thing that is cold. In other words, the thing has remained the same thing, though it is in a different state, at different times in the *A* series. The thing remains permanently part of the *B* series. It is the poker no matter at what point in the series it is fixed. What changes is its state according to its *B*-position. It is the same thing that simply presents different states. These different states however, are not alterations in the thing itself but different events in the history of the one thing. If the thing itself changed then it would not have a continuous history and we could not call it the same thing. If it has all the characteristics of a persisting thing then the only changes that can occur to it, are changes in events that affect, or happen to it. When we say that we have 'changed' over the years we do not really mean that our persistence as a thing has been in any way interrupted. If it had, then we could not say that we were the same people as we were years before. At the same time, we do not generally believe that the material 'thing' that we were when we were, for example, a child, still is the same material thing that we refer to when we refer to our self now; our bodies have probably changed entirely over the years. Although the ideal self survives change, the material self does not. This nature of persistence, and therefore changelessness as a thing, does not rely upon physical sameness but on the persistence of my entity as a spiritual self. The only changes that can happen are in the events that have affected or surround that thing. If we take a materialistic view this description of change is dubious. Here we are dealing with tenseless change, such that *x* is *F* at *t* and *x* is not *F* at t^1 and in this case we have to accept that state of the material thing at *t* is different than its state at t^1. If we take an idealist view it is the thing at the centre of its state — the 'it' in 'its state' that persists and does not change — that we place most important. On this view we can set aside (or downgrade) the tenseless change of the material world and be more correctly concerned with the importance of the question of the persistent spiritual self of the ideal world.

What then precisely is the nature of an event? Common sense tells us that an event is a temporally extended 'something' of which we can be aware of as a whole, for example, a flash of lightning. The flash of lightning however, is very brief. Is an event still an event if it lasts longer than a flash of lightning, say, for example, the duration of an eclipse? Most would agree that an eclipse still is an event, but what about a war? Is that an event, or a lifetime, or the span of the universe? Broad defines an event as, 'anything that endures at all, no matter how long it lasts or whether it be qualitatively alike or qualitatively different at adjacent stages in its history'.[61] Such a definition allows us to feel at ease with events that are as diverse as a flash of lightning and the Second World War. We could, by this definition, include in the category of events, the presently indefinable moment of conception and the equally indefinable total span of a finite universe's existence. Broad, like McTaggart, accepts that there is no such thing as ceasing to exist, there is only ceasing to be present.[62] If an event continues to exist after it has ceased to be present, then it must continue to occur at all moments after its having first been existent. Broad thinks that it is like stuffing 'sausage meat' events into an all accommodating casing. We might say that the unchanging thing is the casing and the sausage meat is the continual addition of events that in *Cambridge change* continue to change the nature of all the other events as they are added.

Marhenke points out something particularly relevant to the views expressed by many about McTaggart's argument. He says, 'McTaggart's argument depends on propositions that we should all unhesitatingly accept, unless we knew in advance that their consequence is the unreality of time'.[63] We do, after all, regard McTaggart's denial of time as a paradox. Because we have an unerring (if confused) belief in the reality of time we therefore tend to believe that though there seems no formal mistake in the argument and the premisses could be true, the conclusion that time is unreal is in error.[64] It may be that the answer to this anomaly lies in the sense in which propositions are understood. Perhaps (though I do not believe so) the propositions that

McTaggart uses cannot be employed as propositions about time while being given a meaning that can then lead to the conclusion that time is unreal. Even if this were the case the fundamental issue of the nature ascribable to events remains. In other words, whether we agree with or argue against McTaggart still we must confront the question of the nature of events. It is important to realise that the question of the nature of events is not an outcome of the denial of the reality of time.

We could suppose that the *B* series may not be a series of permanent terms with permanent ordering relations between them; that events that are 'now' actually exist and because events in the future or past do not there can be no series or order. Here, because the order of succession of terms is permanent would not imply that the terms which succeed one another must be permanent. Mellor applies this argument to the *B* series — which he sees as phenomenal relations — the difference between one event preceding another lying in the *relata* and not in the relation.[65] The same could be applied to the *A* series. The self-contradictory nature of the *A* series could arise from a confusion of types. The idea that the future will exist and the past existed is only ever borne out when they are known as truly existent in the present. Although we should not naturally doubt their reality, such reality can never be demonstrated. But even so, the changing of an event from present to past still would in itself be an event.

An event is something that endures. It may start but never end, or it may start and end in an infinitely short span. Whatever its precise nature is, it remains a classification of something that is happening to an object, whether or not that object is material or spiritual and whether or not it is actually changing or changing in *Cambridge change*. I think the object itself will never alter but the way in which it is classified will. If the poker is hot on a particular Monday that is an event in the history of the poker. If the poker is hot from the start of a particular Monday until the end of the universe (or to infinity), then that is also an event in the history of the poker. This means that there are two different events used to classify the one unchanging poker. The first event of the poker being hot on a particular Monday, changes as it becomes part of the second event that runs to the end of time (or infinity). If the latter event runs to the end of the universe (if there is a last moment) then it becomes the final unchanging event at that last moment. In an infinite universe, the event will continue to change as it changes to infinity.

On the face of it this may seem to conflict with our idea of actual change. After all, we tend to think, in everyday terms, of an event as something that is 'completed', whereas we think of something that still is changing as 'incomplete'. Any thing may have attached to it, events that are either complete or incomplete. This is not incompatible, and we are saved from a serious misconception. This misconception is not relevant in the dynamic system of *Cambridge change*. Here changes in events that effect things continue to infinity or the last moment of the universe.

Events that are transitory can be momentary. Trying to seek out the momentary event can lead us to being caught up in the instantiation of the moment. In looking for the smallest, or shortest moment and therefore the smallest event, we become paralysed by what seems an infinite regress of sub-division. The duration of events can be described in terms of the specious present. However, the human specious present is very different from the specious present of a being for whom the whole of universal existence is just an event. An event is only meaningful if it is a meaningful classification. It is not a quality that applies to meaningless instantiation.

Events are variable in that they are of different 'lengths'. They are also variable according to how we look at them or who we are that looks at them. No specious present could be said to be the same as another specious present. Events are not classifications of occurrences that have any absolute qualities. Each of us views an event in a different way. The determinations of different specious observers must always be different. As Marhenke shows, events are not measurable against any 'universal standard' but determined in different ways by different specious observers.[66]

The specious present of each observer determines the way each event is classified. It is not possible to measure most events with any objective accuracy. Some events can be precisely timed, for example, if I start the clock at 10 seconds past midnight on the 23 February and stop it at 10 seconds past midnight on the 24 February. But mostly, there is more flexibility in the way we describe events and this does not detract from their quality. Events determined in this flexible way are not contradictory. Each specious observer agrees on the vagary of the terms of reference used for classifying the events. For example, if I acknowledge that the Second World War started in 1939 and ended in 1945, that may be acceptable as an appropriately classified event. If necessary, I could agree with someone else, who wanted to be more exact, the month in 1939 and the month in 1945. If pressed, or if we found it necessary, we could agree on an hour of the day in the appropriate month in 1939 and an hour of the day in the appropriate month of 1945. However, even if we did this, still we would have different specious experiences of the event. Particularly if we differed in age or had lived through the period, or had lost a relative in the war and so on. What we could not do is find a precise moment at which the Second World War started and a precise moment at which it ended. We can therefore never absolutely define the event. In addition, we would also have to accept that in *Cambridge change* we could never fully classify that event until all the facts were in. This would include all historical facts that continue to become known as well as all facts remaining in the universe. In this way the event of the Second World War is not complete until all subsequent events are known and perhaps this process would continue to infinity. It would also have to include all the specious experiences of that event, as well as all events in the *B* series that were earlier than it. It will therefore never be capable of complete description.

This is similar for material things. For example, where does Mount Everest start or end or where does the pattern of a carpet start or finish? The difference is, however, that the classification of such objects is different from the classification of a temporal event. We can talk about 'how long Mount Everest has endured' or 'how long the pattern has been on the carpet', but to talk about where Mount Everest starts or ends or where the pattern of the carpet starts or finishes has no *A* or *B* series significance. The size of Mount Everest or the limits of the pattern of the carpet are not earlier or later than anything, nor are they in the past, present or future of anything. They are part of the metric of the material world. Again this shows how concentration on changes in material objects can confuse not only how we judge the nature but also the temporal significance of events.

If events are what make up the nature of change, then any apparent changes in things are misperceptions. To accept this, however, we need to reject the material world that seems such a certain part of our everyday experience. Doing this is no loss in searching for the real meaning behind time which has little coherence as part of our everyday experience.

3.4 The vicious regress

In this section I show that the argument for the vicious regress of the *A* series is not so easily answered as some think. To do this, in §§ 3.4.1, 3.4.2, 3.4.3 and 3.4.4, I analyse four arguments commonly used to attack it.

The notion of events changing leads naturally to the problem of what it is that events change in. I think the point has been made that it is unlikely that events can change in the *B* series so we are led back to the *A* series and the contradiction in its terms that results in the vicious infinite regress. When considering the regress there are three alternatives. 1. That the *A* series gives rise to a vicious infinite regress. 2. That it is a mistake in the first place to use

the *A* series category to represent some kind of change which events undergo.
3. That the vicious regress is irrelevant to the argument.

McTaggart argues for '1', that the *A* series gives rise to a vicious infinite regress and that this closes the circle which encompasses its unreality. I shall return to this after disposing of both '2' and '3'.

Alternative '2', that it is a mistake in the first place to use the *A* series to represent change in events, can be set aside. We have seen that the *B* series cannot sustain change and therefore any change that occurs can only be found in the *A* series.

Alternative '3', that the vicious regress is irrelevant to the argument. If the contradictory nature of the *A* series is, as Geach puts it, bogus, there is no point in pursuing whether any attempt to remove the contradiction leads to a vicious regress. Geach thinks that Wittgenstein is right in saying that when philosophers think that the problem is a vicious regress, the problem really lies at the first step.[67] Geach says, 'if McTaggart's alleged contradiction is bogus, we need not consider whether the attempt to remove it would generate a vicious regress; if it is not bogus, it suffices to show that *A*-characteristics are not instantiated'.[68] If the *A* series contradiction is correct still we can feel that there is no point in pursuing it, simply accepting that *A*-characteristics are then 'not instantiated' and that is all there is to it. This, however, is unsettling. The contradiction in the *A* series is not completely shown unless the vicious regress is involved. We simply do not know that the separate phases of change in the *A* series are incompatible unless that incompatibility is seen in the regressive nature of the terms. This point is unavoidable.

Now point '1', that the *A* series gives rise to a vicious infinite regress. This is crucial, for if we accept the *A* series and the vicious infinite regress is upheld then time cannot be real. There are many arguments against the vicious regress. Some have argued that the regress although a regress is not vicious, some that it is not infinite and others that it is not infinitely vicious. Though it is more popular these days to subscribe to a strictly *B* series view and thus ignore the *A* series (and so the vicious regress) altogether, there are some that take up the matter. As examples, Sandford, responding to Gale's assertion that McTaggart's argument is 'fallacious in ... a deep and basic way',[69] and supporting Gotshalk who asserts that the apparently self-contradictory nature of the *A* series is untrue,[70] attacks the contradiction implicit in the *A* series.[71]

How we look at the contradiction implicit in the *A* series, bears upon which of McTaggart's versions of the denial we read. In Chapter 33 of *The Nature of Existence* McTaggart talks of a 'vicious infinity' and in the 1908 article in *Mind* of the 'vicious circle'. If it is a vicious circle — a *reductio ad absurdum* — (a principle first proposed by Poincaré in 1906 in an attempt to diagnose some of the contradictions of logicism), Sandford thinks McTaggart has to presuppose the *A* series to account for the *A* series. Sandford says this circularity is just 'explaining away the alleged self-contradictory nature of the *A*-series'.[72] The contradiction, according to Sandford, follows from the assumption that there *is* an *A* series, but this assumption, from which the contradiction is said to follow, cannot then be disallowed in an attempt to show that it does not really follow. Here one *A* series needs another *A* series in which to happen and that *A* series needs another *A* series and so on. This need not be, as Mink calls it, a meta or meta-meta-time.[73] There is no reason we cannot say, for example, that a moment is past at a future time without introducing a meta-time. To suppose a moment as past in a future time in the *A* series requires a second *A* series and so on and this is a vicious circle. But there is no circularity in the notion of the vicious infinity which does not require a meta-time, or second or higher order time series. This is resolved in *The Nature of Existence* where McTaggart no longer talks about 'vicious circularity'. More appropriately then, we choose the argument of the vicious infinite series.

In *The Nature of Existence*, when McTaggart talks of the 'vicious infinity', he is not talking about an infinity of different time series but an infinity of different moments that occur in the same time series. Allowing that the time series itself is, however, only one series (and immune from vicious circularity) the moments that occur in the series are contradictory when viewed successively. No moment can be past and future and present; this could well be a 'vicious truth'.[74]

McTaggart says that present events are inconsistent. If *M* is present, there is no moment of past time at which it is past. But the moments of future time, in which it is past, are equally moments of past time, in which it cannot be past.[75] But this can be re-written. If *M* is present, then there is no moment *N* such that *N* is past at a moment that is present, and *M* is past at *N*. If *M* is present, then for any moment *N* that is future at a moment that is present and at which *M* is past, there is a moment *O* such that *N* is past at *O* in which *O* need not be a moment that is present. This removes the contradiction.[76]

The same can be done for McTaggart's statement on the inconsistency of future events which he says 'cannot be present or past at any moments of past time. But all the moments of future time, in which *M* will be present or past, are equally moments of past time'.[77] This can be re-written. If *M* is future, then there is no moment *N* such that *N* is past at a moment that is present, and *M* is present at *N* or *M* is past at *N*. If *M* is future, then for any moment *N* that is future at a moment that is present and at which *M* is present or past, there is a moment *O* such that *N* is past at *O*. In which again *O* cannot be a moment that is present.[78] Sandford thinks that this re-writing answers McTaggart's problem of contradiction in the *A* series. Though, he concedes this in itself does not support the reality of time, only that McTaggart's argument for the contradictory nature of the *A* series is invalid. Indeed, he sensibly holds back from intuitively standing against McTaggart's conclusion, warning of the arguments of philosophers with 'axes to grind' that argue against McTaggart more because of his atemporal stance than anything else.

We can, like Christensen, also make a tense-logical attack on McTaggart's misuse of *A*-predicates and the consequent invalidity of the *A* series;[79] an attack, as Christensen acknowledges, first given quite clearly in Broad's *Examination*.[80] In doing this we would accuse McTaggart of modifying 'past', 'present' and 'future' by adding 'is' to make 'is past', 'is present' and 'is future' (or 'was' or 'will be') and declaring that the 'is' is tenseless so causing contradiction. By taking this 'ordinary language' view against McTaggart we could assert that 'is' is tensed and cannot be used in any other way.[81] This view holds that there is no regress because all that is happening is McTaggart is trying to get rid of the tense in 'was' (as in 'was future') by replacing it with 'at a past time'. This leads to a further tensed statement because 'at a past time' means 'at a time that is now past'. McTaggart is thus forced into further steps in the regress to de-tense the *A*-predicates in each added propositional phrase. Thus, 'Every succeeding stage of the regress is generated, not by the defender of time trying to escape a contradiction, but by McTaggart trying to produce one'.[82] The argument goes that what McTaggart thinks of as propositional phrases are adverbial in nature and that there is a mistake in believing that an adverb with a complex inner structure, such as 'at a moment of past time', is more basic than more basic structureless tenses (Christensen calls these '*A*-adverbs'[83]). Thus, McTaggart is in error by attempting to reduce the *A*-adverbs to *A*-predicates via the *A*-predicates that appear in the propositional phrases. On the other hand, ordinary language will not allow the reduction of tenses to *A* predicates in this way. There is no apparent sense in saying a moment 'be past' as opposed to saying 'will be past'. In ordinary language, *A*-adverbs are logically more basic than *A*-predicates. For example, 'event *e* occurred' contains the verb 'occur' but no temporal predicates, '*e* occurred' is equivalent in meaning to '*e* is past' and '*e* was present' and these contain the *A*-predicates 'is past' and 'was present'. The '*A*-names' 'past', 'present' and 'future' are deniable in the same way. *A*-predicates and *A*-names can only appear if accompanied by *A*-adverbs. Therefore, of these three logically related terms, *A*-adverbs are the fundamental ones. Prior, who says Geach made him aware of important positive aspects of McTaggart's work,[84] produces a logical system that he claims makes exact the study of temporal adverbs.[85]

In the *A*-adverbial world, 'is past', 'is present', 'is future' do not really signify anything that is a real attribute or quality of anything. The idea that an object could acquire an everlasting quality simply by occurring or ceasing to exist is unacceptable. Christensen says this sort of existence, 'is merely one which did exist but does not exist'.[86] On this account events that are either past or present are not special kinds of events at all. The tenses from which we derive such predicates only tell us when a thing had those properties. They only tell us when it had such characteristics, whether it does now, or whether it will have them in the future. More generally, because the sort of information that tenses carry is so unusual and takes on a very particular form, the concept of time is very different from all other concepts. *A*-adverbs need great precision of definition and this often leads to abstraction of a high order. Whereas Christensen does not get involved in 'now', Grünbaum defends 'nowness' as a special kind of complex mental event. He refers to 'the adverbial attribute now' and treats this *A*-adverb as a sort of attribute-designating predicate.[87]

McTaggart is, in this way, accused of being lulled into a comfortable attitude that allows him to think that tenses can be replaced by predicates because of our deep-seated reliance upon the subject-predicate language form. Such a monadic form bears little resemblance to the form of tense-logic. The problem, it is said, is compounded because our belief in the language system we use causes us to develop concepts concerning the properties and relations of existents. This leads to hypostatizing abstractions, not least of which is the term 'time'. Because time does not fit comfortably into our everyday language it appears to us enigmatic. It is therefore mistakenly thinking in these terms about time that causes us problems and such thinking will never lead us to a full understanding of its nature.

This attack, though, is ill-founded. Whereas it accuses McTaggart of an intuitive (and premature) hypostatizing of time as relying on the *A* series it commits a greater offence by hypostatizing the components of the *A* series in order to show that the *A* series is misconceived. That it does this under the guise of making 'ordinary language' sense of what it considers easily abstracted temporal terms only compounds the error. There is little to be gained from further analyzing the components of the *A* series in an attempt to reveal their falsehood as this only begins an internally recurring falsehood that sits inside the *A* series vicious regress.

It is a common idealist view that the *B* series describes 'real' time and the *A* series is subjective. McTaggart's view, though supporting this basic standpoint, leads us to find the *B* series incapable of supporting time and therefore incapable of revealing 'real' time. The *A* series, which may be subjective, is the fundamental series and change can only occur in the characteristics of events in the *A* series. Furthermore, when we look closer, we find that the terms of the *A* series are contradictory. McTaggart thinks that the *B* series is the only place where time can be found. However, without an *A* series it is not real. Because the *A* series is contradictory the support for the *B* series disappears and so does its reality. Whatever the *B* series is, it is not 'real' time.

McTaggart's argument shows that the *B* series is incapable of sustaining time. That change, as *Cambridge change*, occurs to events, so that our 'normal' view of temporality is not necessary. However, if the *A* series is valid it would provide so much support for the *B* series that the idea of time being unreal would be difficult to hold. To show that the contradiction in the *A* series really does amount to a vicious infinite regress would put the *A* series to flight. To show that this cannot be done I now look at arguments that are commonly used to attack the *A* series regress.

3.4.1 A-terms are used at different times

This argument says we use the terms of the *A* series at different times. We say an event is present *or* past *or* future at a given moment or present *and* past *and* future in successive moments. It is only in this latter case (a variant used by Gunn and later taken up by Mellor), where we say it is present *and* past *and* future in successive moments, that the contradiction arises.[88]

I do not think this alternative (the 'hydra-headed' nature of the *A* series, as McTaggart calls it) gets rid of the contradiction. It is not acceptable that we make recourse to a succession outside either the *A* or *B* series. If the succession is only in the *B* series then there can be no change. All 'succession' would be fixed in a static universe, being always, and having been, in strict relation of earlier or later. If the succession is in the *A* series then the characteristics of the terms of the succession change, but only in relation to another *A* series and so the regress (or at least a vicious circle) starts again. In order for change to be real then it must occur in *A*-succession.

3.4.2 A-terms used to describe A-terms are equivalent terms

This argument has strong supporters including Wisdom,[89] Stebbing, Cleugh and Mellor. It claims it is wrong to conclude from the statement 'something is present at a moment that is present' that this leads to a further statement that 'something is present at a moment that is present at a moment that is present', which then leads to the statement that 'something is present at a moment that is present at a moment that is present at a moment that is present' and so on. The statements are equivalent in meaning; the second statement is consequent on the first and the third consequent on the second and the first, and so on but still they are equivalent and no vicious regress is forthcoming.

But such statements are not equivalent in meaning because one entails the other. 'Agnes is Harry's spouse', entails 'Agnes is Harry's spouse's spouse's spouse', entails 'Agnes is Harry's spouse's spouse's spouse's spouse's spouse', and so on (Mink's example).[90] Such an entailment process is indeed not vicious because the statements that follow the first, although consequent on the first, are not a condition of the truth of the first. Such a consequential set of statements cannot lead to a vicious regress because each subsequent statement does not rely upon the previous statement to be true. Only if such reliance was present would the conditional nature of each subsequent statement lead us into a vicious regress. Such statements do not parallel the *A* series regress (in which the terms are equivalent and viciously regressive) and so cannot be used to prove it wrong.

Also, the argument is usually demonstrated using examples that only refer to the present (Stebbing and Wisdom in particular do this) so that repetition of 'is', which refers only to the present, is redundant. Whereas, 'was' and 'will be' imply a present that can only be made explicit by further analysis which involves equivalence and the regress. Also, there is a difference between an event, and moment, being present and an event, and moment, being past. If we can accept saying 'The Second World War is past', while at the same time not accept saying '1945 is past' then we can go along with this. However, if we find meaning in saying '1945 is past' then we can disregard the argument, for it does not allow us a method of analyzing *A*-predicates. Wisdom recognises this, but says that 'is', 'was' and 'will be' are not capable of analysis as each is some kind of fundamental fact.[91] But there is nothing in this view that compels us to accept a notion of the irreducibility of tenses. The present tense may be incapable of analysis but this cannot apply to future and past as well. If we accept such irreducibility not only must it apply to the simple monadic tenses like 'I was' but to triadic tenses like 'I was about to have been'. Indeed, there may be *n*-adic tenses of any degree, all of which, on this argument, would be equally irreducible as 'I was', 'is' and 'will be'[92] and this is a highly unlikely proposal. It makes little sense to confine any argument against the vicious regress of the *A* series to the present tense when it only arises in relation of the present to either the future or the past.

3.4.3 A B-language is all that is necessary

This argument focuses on the claim that McTaggart uses language to corrupt the meaning of tense. He is often accused of this, and in his argument against time notably by Pears, Smart and Findlay. They accuse him of creating complex problems with language by de-temporalizing ordinary verbs and yet expecting of them the same grammatical logic as when temporalized. If instead (as Findlay thinks[93]), we restrict ourselves to the use of 'dates' and 'tenseless participles' then the difficulties that we have with the *A* and *B* series do not arise. But to use *A*-characteristics and *B*-characteristics without them in some way presupposing one another, seems to propose a linguistic perplexity that would be even more confusing.[94] As we have already seen a pure *B*-language is impossible because 'earlier' and 'later' have no meaning unless we can denote a 'present' which could not form part of a *B*-language.

The *A* series still may be, as Gale calls it, 'impure'.[95] McTaggart believes that if time consisted in only a *B* series, though there would be different possibilities for change none of them would make sense. Without an *A* series *Cambridge change* is no longer possible and the *B*-world is reduced to a permanently fixed series of events. The positional quality of events makes redundant any alteration to events themselves which we normally think of as change. The idea that in a *B*-world the death of Queen Anne could also cease to be the death of Queen Anne is absurd. Whatever was the permanently positioned event must always remain the permanently positioned event. Two events could merge to become one event for each event would have to cease to be what it was originally. It would be logically impossible for any event to escape from the *B* series; the *B* series is a permanent state and nothing can 'get out' of it once it is 'in'. Events could not change their position in a *B*-world; the permanent relation that *B*-terms have to one another is unchangeable, if *N* is earlier than *P* then it will always be earlier than *P*.

Because the *B* series alone cannot sustain change, McTaggart thinks that the only place where change can occur must be in the *A* series where a change in the position in the series can be involved. But are we entitled to make such a demarcation between what is to do with the *A* series and what is to do with the *B* series?

To return to Russell's objection that it is things and not events that change: that 'the poker is hot on Monday' and 'the poker is not hot at times other than Monday'.[96] McTaggart's argument is that both these qualities are true of the poker both when it was hot and when it was not hot and therefore there has been no change in the poker. This is because the poker is 'always' hot on that Monday and 'always' not hot at any other time because of its permanent position in the *B* series. But the tense-logical view says that 'always' is a tenseless operator and is wrongly included in a tenseless logic.[97] The obvious response is to say that the claim that things change does not require that tenseless truths about those things change; rather, change in things is analysed in terms of the different tenseless truths, mentioning different *B* series positions, which apply to things. We cannot make sense, however, of *B* series positions independently of the *A* series, in our experience of time. Mellor would (I think) respond to this phenomenological point by employing his token reflexive analysis of the truth conditions of tensed sentences but we have already seen that this invokes non-equivalent *A*-terms.

There is an asymmetry in the relation between *A*-language and *B*-language. Anything that we can say in *B*-language we can say in *A*-language but not the other way around. From my present position in the *A* series I can talk about 'earlier' and 'later' events in the past and 'earlier' and 'later' events in the future. If these events in the future become events in the past the *B*-relations remain permanently fixed. However, if I adopt a *B*-relation position, although all events in the future are 'later than' events in the past, when they move into the past they do not remain 'later than' but become 'earlier than' any subsequent future event. Gale concedes this point, that the *B* series cannot be reduced to the *A* series even though, 'there is an asymmetry in information content between a pure *B*-relation language and a pure *A*-deterministic language, in that everything sayable in the former is sayable in the latter but not vice-versa'.[98] This approach only throws us back onto some form of *A*-language and once this is re-introduced the argument for a *B*-language only dissolves.

Tense-logic will only allow that the *B* series can be reduced to an 'impure' *A* series. That is an *A* series that is determined by itself and *B*-determinations, first by past, present and future and then by more past and more future. As Gale says, 'an *A*-Series which is determined not only by an unqualified past, present, and future but also by more past and more future'.[99] Grünbaum also favours this non-metrical concept of overlapping temporal intervals that can be described as 'more past' or 'more future'.[100] But any such impure *A* series is also a *B* series.[101] The generating relation of the impure *A* series is therefore 'earlier than', which is also the generating relation of the *B* series.

The impure *A* series is essentially two series, one of 'more past than' and another of 'more future than' both of which 'join together' at two of their extreme edges (*terminus ad quem* and *terminus ab quo*) at whatever is the present event. Using the 'more' future and 'more' past concept (as Gale acknowledges[102]) raises a plausible objection though. These statements involve a tenseless *B*-relation in that if '*x* is now more past than *y*' (an impure *A*-relation statement involving the 'more past' concept) this is analyzable into the *B*-relation statement '*x* is earlier than *y*' and the *A*-determinative statement '*y* is now past'. If this is true then the situation is circular and provides no escape from the apparent fact that reducing *B*-relations to *A*-determinations contains *B*-relation statements.

The statement '*x* is more past than *y*', however, is only an impure *A*-determinative (or tensed) statement because it has the same restrictions placed upon it as the (pure) *A*-determinative statement '*x* is now past'. The statement '*x* is more past than *y*' made at different times causes it to have different truth-values that will not happen to the *B*-relation statement '*x* is earlier than *y*'. Nevertheless, this does not prevent us from analyzing '*x* is more past than *y*' into a conjunction of a *B*-relation statement and an *A*-determinative statement the outcome of which will always be an *A*-determinative statement. Instead of analyzing '*x* is more past than *y*' into the conjunction of '*x* is earlier than *y*' and '*y* is now past' we could say that '*x* is more past than *y*' entails '*x* is earlier than *y*'.

If we use, for example, 'preceded' instead of 'more past than' or 'before' or 'while' instead of 'earlier than' or 'simultaneous with' we reveal circularity. Even so, if we use 'more past than' and 'more future than' statements we show that the *B* series cannot be reduced to a pure *A* series. It remains that our concept of time involves the three unanalyzable elements of the pure *A* series, past, present and future and a separate set of problems regarding our idea of ordering those elements. In other words, our concepts of 'earlier than' and 'later than' in the *B* series.[103]

Yet, as far as the *B*-theorist is concerned, the fact that one event will always be either earlier or later than another event is enough to describe the world. But this requires that there are at least two events in the universe and that may not be the case; an event of the whole span of the universe is just as much an event as any 'smaller' specious event. It also ignores our need for a realistic interpretation of change. If event *M* is earlier and event *N* is later it would not be possible to form a concept of time because nothing can change. We have a *B* series but nothing temporal has happened. In supposing a *B*-language only we are ignoring change and if we have at least two events we cannot do that if we are to have some notion of time. There must have been some change taking place between *M* occurring and *N* occurring. Even if *M* and *N* related to the same object they would not be separately *M* and *N* if some change had not occurred. To try to support a temporal concept without using temporal terms, far from removing difficulties, places more difficulties in the way. It requires us to attach a catalogue of dating labels onto events in the *B* series. This would involve us knowing in what order those 'date labels' were in respect of to make any sense of them. Added to which, if we did this we would be supporting the *B* series with a list of *A*-features and still would not have escaped the temporalization of the events in question.

We cannot correctly assume that *M* was in fact earlier than *N* without any temporal terms. We cannot say that *M* was earlier than *N* unless we can relate it to something else. If we only use *B*-relations then all we can conclude is that *M* was one event and *N* was another event. To say that one event was earlier than the other, without additional reference, means nothing temporal. We cannot say one event was earlier than another event unless we know 'where' at least one event was in relation to another set of terms. These terms can only be found in the *A* series.

3.4.4 The A series is all that is necessary

This argument proposes that even if the *B* series is not temporally valid without the *A* series, the *A* series is temporally valid without the *B* series. If this were so then such an *A* series might not involve the original contradiction and the vicious regress would be avoided.

But a pure *A*-language is also impossible. Once the *B*-relations are removed, we are left only with terms that describe a unique tense. Past events, if they cannot be seen as earlier than something, mean nothing other than that they are past. Future events, unless we know what they are later than, mean nothing other than that they are future. The *A* series is meaningless as a time series unless we can put it into the temporal context of the *B* series.

Mink suggests that we could try to assemble a temporal order by using a series of *A* series.[104] In such a series each moment implies a particular *A* series, the preceding series to which is now 'more past' and a previously related future event is now 'less future'. But this does not work. 'More past' and 'less future' are *B*-features that help us locate *A*-features. To get round this we would require as many *A* series as there are moments. Not only would this be cumbersome and impracticable but to mean anything temporal still it would have to be interpreted as a *B* series.

3.4.5 The B series is all that is necessary

Another idea is that we could think of the world-process as a *B*-model manifold series of events that represents itself. In Mink's terms, 'strung out along a dimension of occurrences and jointly constituting the world of fact'.[105] In this model we have to accept, as included, that we view time, memory and the specious present in a spatial way. Dates are convenient and useful in the model but can equally be set aside. We would have to learn to be at ease with a chronology, or more properly an 'ordering', which operated entirely on 'earlier' and 'later'. Dates are after all only abbreviations for long strands of *B*-relation explanations. In such a world I could describe 1945 either by saying '1945' or by beginning, '1945 was earlier than the first event in 1946, and that event was earlier than the second event in 1946' and so on. But this falls into the same trap as the *B*-language only model. Without true *A*-features in support it just does not make sense. It is not only unacceptable in everyday terms but it conflicts with any real 'sense' of time that we may have.

3.4.6 Consciousness can replace the A series

We could suppose that the apparent movement in time is caused by our consciousness as everything squeezes through the thin (or thick) gateway of our specious present. In this world the future and the past are only possibilities, let in through our consciousness, of some kind of temporal flow. But even if we can attach any meaning to such a 'present consciousness', we are unavoidably caught up in the idea of the 'temporal flow' and therefore an *A* series. Pushing it to the sidelines does not make it go away. Adopting the *B* series only model means we cannot say what we mean about time; adopting the consciousness-model we can never be sure that we mean what we say.

I conclude that we must accept that although, on the face of it, the vicious regress seems bogus, common arguments used to answer it are not convincing.

3.5 Conclusion

In this Chapter we have covered some much disputed ground. In §§ 3.2.1-3.2.3 we saw how McTaggart divides time into two series (*A* and *B*) and using this classification, provides an argument for denying its reality. We then saw that accepting this presents us with psychological problems. In § 3.2.4 we found that our experience of change as we know it can no longer be held. In § 3.2.5 we saw that our experience of time in the *B* series must also be wrong. In addition, we also concluded that denying the reality of the *B* series does not provide an alternative to time because still it is denying time as we know it. This makes many arguments against McTaggart's denial, although strongly made, unsuccessful in re-establishing an acceptable temporality. In § 3.3 we saw that if we accept that it is events that make up the nature of change, then any apparent changes in things are misperceptions. However, to adopt this view demands that we hold a strong idealist view and expects us to reject much of what seems part of our experience. In § 3.4 we saw that conventional attacks on the vicious regress are not altogether convincing. This provides strong (if not convincing) support for McTaggart's argument against time. After all, on McTaggart's argument, if the vicious regress is true, time is logically contradictory. In this way, we have arrived at some reasonable grounds for accepting atemporality, although common sense tells us otherwise.

I acknowledge that it is difficult to accept that it is events and not things that change — we have to take a strong idealist view to accept this — however, defence of the vicious regress can be found. McTaggart's argument against time provides us with enough suspicion about the reality of time to encourage us to see what it is he replaces it with. In doing this we take a decisive step forward from nearly ninety years of inconclusive argument about McTaggart's paradox.

Notes

1 D.W. Gotshalk, 'McTaggart on Time', *Mind*, vol. 39 (1930), no. 153, pp. 26-42.
2 McTaggart, *The Nature of Existence*, § 303.
3 Ibid., § 304.
4 Ibid., § 304.
5 Ibid., § 350.

6 Geach, *Truth, Love and Immortality*, p. 89.

7 Aristotle, *Physica (Physics)*, trans. by R.P. Hardie and R.K. Gaye in *The Basic Works of Aristotle*, ed. and with an Introduction by R. McKeon (New York, Random House, 1941) Bk. 5, ch 1, 236^{b}, line 20.

8 Ibid., 237^{a}, lines 17-20.

9 For an alternative view see S. Shoemaker, 'Time Without Change', *Journal of Philosophy*, vol. 66 (1969), no. 12, pp. 363-81.

10 See C.D. Broad, *Scientific Thought* (London, Kegan Paul, Trench, Trubner and Co., 1923), pp. 282-86. *Scientific Thought* contains criticism of McTaggart's theory and McTaggart was able to respond to Broad's criticism when he wrote his chapter on time in *The Nature of Existence*. McTaggart, who held Broad in the highest philosophical esteem, though at the same time acknowledging him as deeply opposed to his own ideas, took Broad's criticism and his ideas about time seriously.

11 McTaggart, *The Nature of Existence*, § 307.

12 Ibid., § 308.

13 Welby, 'Mr McTaggart on the "Unreality of Time"'.

14 McTaggart, *The Nature of Existence*, § 311.

15 Ibid., § 332.

16 Ibid., § 327.

17 Gotshalk, 'McTaggart on Time', p. 33.

18 R.M. Gale, 'McTaggart's Analysis of Time', *American Philosophical Quarterly*, vol. 3 (1966), no. 2, p. 150.

19 A. Grünbaum, *Philosophical Problems of Space and Time* (New York, Knopf, 1963), ch. 1.

20 McTaggart, *The Nature of Existence*, § 327.

21 Ibid., § 328.

22 Ibid., § 329.

23 This protection against future criticism was added by McTaggart as a footnote to *The Nature of Existence*, § 332 in Draft *C* of the typescript. It shows in some part a concern for the side-stepping of an appropriate definition of the terms of the *A* series in relation to the 'external term'.

24 Ibid., § 331.

25 Ibid.

26 Ibid., § 332.

27 Ibid., § 333.

28 M. Dummett, 'A Defence of McTaggart's Proof for the Unreality of Time', *Philosophical Review*, vol. 69 (1960), no. 392, pp. 497-504.

29 Geach, *Truth, Love and Immortality*, p. 102.

30 Ibid., pp. 90-93.
31 McTaggart, *The Nature of Existence*, § 309.
32 Geach, *Truth, Love and Immortality*, p. 91.
33 Farmer, *Being in Time*, p. 20.
34 S.V. Keeling, *Time and Duration*, ed. G. Rochelle with an Introduction by E. Senior (Lewiston, New York, Edwin Mellen, 1991), pp. 48-9.
35 G.J. Whitrow, *Time in History* (Oxford University Press, 1989), p. 172.
36 See H. Bergson, *Creative Evolution*, trans. by A. Mitchell (London, Macmillan, 1911).
37 Geach, *Truth, Love and Immortality*, p. 91.
38 For a modern account of the subjectivist or consciousness related nature of the *A* series determinations see Mellor, *Real Time*.
39 H.D. Oakeley, 'Time and the Self in McTaggart's System', *Mind*, vol. 39 (1930), no. 154, pp. 175-93.
40 McTaggart, *The Nature of Existence*, § 313.
41 B. Russell, *The Principles of Mathematics*, 2nd edn. (London, George Allen and Unwin, 1985. First Published 1903), p. 469.
42 McTaggart, *The Nature of Existence*, § 314.
43 Ibid., § 315.
44 Ibid., § 318.
45 Broad, *Examination*, vol. 2, pt. 1, p. 296.
46 L.S. Stebbing, 'Some Ambiguities in Discussions Concerning Time', *Philosophy in History: Essays Presented to Ernst Carrirer*, ed. R. Klibansky and H.J. Paton (Oxford, Clarendon Press, 1936), pp. 107-23.
47 M.F. Cleugh, *Time and its Importance in Modern Thought* (London, Methuen, 1937), pp. 148-65.
48 Marhenke, 'McTaggart's Analysis of Time', *University of California Publications in Philosophy*, vol. 18 (1935), pp. 15-74.
49 D.F. Pears, 'Time, Truth and Inference', *Proceedings of the Aristotelian Society*, vol. 51 (1950-51), pp. 1-24.
50 J.J.C. Smart, 'The River of Time', *Mind*, vol. 58 (1949), no. 232, pp. 483-94.
51 Geach, *Truth, Love and Immortality*, p. 95.
52 L.O. Mink, 'Time, McTaggart and Pickwickian Language', *The Philosophical Quarterly*, vol. 10 (1960), no. 40, pp. 252-63.
53 A.N. Prior, 'Changes in Events and Changes in Things', in R. Le Poidevin and M. MacBeath, eds., *The Philosophy of Time* (Oxford University Press, 1993), pp. 35-46.

54 Ibid., p. 36.
55 Ibid., p. 38.
56 Ibid., p. 44.
57 H. Bergson, *Time and Free-will*, trans. by F.L. Pogson (London, Allen and Unwin, 1910).
58 M. Merleau-Ponty, *The Phenomenology of Perception*, trans. by C. Smith (London, Routledge and Kegan Paul, 1962).
59 A. Grünbaum, 'The Status of Temporal Becoming', in R.M. Gale, ed., *The Philosophy of Time* (London, Macmillan, 1968), pp. 322-54.
60 J.J.C. Smart, *Philosophy and Scientific Realism* (London, Routledge and Kegan Paul, 1963).
61 Broad, *Scientific Thought*, p. 54.
62 Ibid., p. 69.
63 Marhenke, 'McTaggart's Analysis of Time', p. 152.
64 Somewhat like Priestley who recalls that though convinced by McTaggart's arguments in lectures, he never believed them afterwards. (J.B. Priestley, *Man and Time* (London, W.H. Allen, 1978. First published 1964), p. 45).
65 D.H. Mellor, *Real Time* (Cambridge University Press, 1987. First published 1981), p. 143.
66 Marhenke, 'McTaggart's Analysis of Time', p. 160.
67 Geach, *Truth, Love and Immortality*, p. 100.
68 Ibid., p. 101.
69 R.M. Gale, 'The Static Versus the Dynamic Temporal', in Gale, *The Philosophy of Time*, pp. 65-85, p. 65.
70 Gotshalk, 'McTaggart on Time'.
71 D. Sandford, 'McTaggart on Time', *Philosophy*, vol. 43 (1968), no. 166, pp. 371-78.
72 Sandford, 'McTaggart on Time', p. 373.
73 Mink, 'Time, McTaggart and Pickwickian Language', p. 253.
74 Sandford, 'McTaggart on Time', p. 375.
75 McTaggart, *The Nature of Existence*, § 331.
76 Sandford, 'McTaggart on Time', pp. 377-8.
77 McTaggart, *The Nature of Existence*, § 331.
78 Sandford, 'McTaggart on Time', p. 378.
79 F. Christensen, 'McTaggart's Paradox and the Nature of Time', *The Philosophical Quarterly*, vol. 24 (1974), no. 97, pp. 289-99.
80 Christensen's own reference is to the portion of Broad reproduced in Gale, *The Philosophy of Time*, pp. 117-42, this being an abbreviated version of ch.

35, 'Ostensible Temporality', in Broad, *Examination*, vol. 2, pt. 1, pp. 264-323.

81 Christensen, 'McTaggart's Paradox and the Nature of Time', p. 291.

82 Ibid., p. 291.

83 Ibid., p. 294.

84 A.N. Prior, *Past, Present and Future* (Oxford, Clarendon Press, 1967), p. vi.

85 A.N. Prior, 'Postulates for Tense-Logic', *American Philosophical Quarterly*, vol. 3 (1966), no. 2, pp. 153-61.

86 Christensen, 'McTaggart's Paradox and the Nature of Time', p. 297.

87 A. Grünbaum, 'The Meaning of Time', in N. Rescher et al., eds., *Essays in Honor of Carl G. Hempel* (Dordrecht, 1969), pp. 147-77.

88 J.A. Gunn, *The Problem of Time. An Historical and Critical Study* (London, George Allen and Unwin, 1929), p. 348.

89 J. Wisdom, 'Time, Fact and Substance', *Proceedings of the Aristotelian Society*, vol. 29 (1928-29), pp. 69-94.

90 Mink, 'Time, McTaggart and Pickwickian Language', p. 256.

91 Wisdom, 'Time, Fact and Substance', p. 92.

92 See H. Reichenbach, *Elements of Symbolic Logic* (New York, Macmillan, 1947), pp. 287-98.

93 J.N. Findlay, 'Time: A Treatment of Some Puzzles', in A.G.N. Flew, ed., *Essays on Logic and Language* (Oxford, Basil Blackwell, 1951) pp. 37-54, p. 53.

94 Mink, 'Time, McTaggart and Pickwickian Language', p. 259.

95 R.M. Gale, 'McTaggart's Analysis of Time', *American Philosophical Quarterly*, vol. 3 (1966), no. 2, pp. 145-52.

96 B. Russell, *The Principles of Mathematics*, p. 472.

97 Gale, 'McTaggart's Analysis of Time', p. 148.

98 Ibid., p. 149.

99 Ibid., p. 150.

100 Grünbaum, *Philosophical Problems of Space and Time*, ch. 1.

101 Gale, 'McTaggart's Analysis of Time', pp. 150-51.

102 Ibid., p. 151.

103 For a full appreciation of Gale's account of the logic of tensed and tenseless statements see: R.M. Gale, 'Endorsing Predictions', *The Philosophical Review*, vol. 70 (1961), no. 395, pp. 376-85; 'Tensed Statements', *The Philosophical Quarterly*, vol. 12 (1962), no. 46, pp. 53-9; 'A Reply to Smart, Mayo, and Thalberg on "Tensed Statements"', *The Philosophical Quarterly*, vol. 13 (1963), no. 53, pp. 351-6; 'Is It Now Now?', *Mind*, vol.

73 (1964), no. 289, pp. 97-105 and 'Existence, Tense, and Presupposition', *The Monist*, vol. 50 (1966), no. 1, pp. 98-108.

104 Mink, 'Time, McTaggart and Pickwickian Language', p. 260.

105 Ibid., p. 261.

4 Time and the *A* Series

4.1 Introduction

In the previous chapter we saw that McTaggart's argument can stand against claims that it is unacceptable. We know that the outcome of his denial of the reality of time leads him to believe that reality is to be found in an atemporal world. But we also know that his argument about time is paradoxical; it shows that something which seems undeniably the case is not the case. If we continue to see it like this, we will continue to intuitively reject his claim no matter how rationally plausible it might seem. I shall now show that any intuitive claim we have for accepting the reality of the *A* series is wrong and that none of our commonly held views on time in the *A* series are fully coherent. Doing this will remove the foundations for our intuitive belief in its reality and remove the 'paradox' from McTaggart's conclusion that it is unreal.

In §§ 4.2 I look at what I consider all the major views that either influence our opinions about time or to which we have recourse when we need some answer about its nature. I show that it is part of the human condition to be concerned with the nature of time. I point out that there are a number of philosophical views to be taken on time. These fall into two main categories: absolute time and relational time. I describe how science has taken on much of the responsibility for answering questions about temporality. I also show how the idea of the non-existence of time poses questions that cannot be answered using conventional views. I describe how common sense is an important feature when dealing with the apparently paradoxical claim on the unreality of time. According to McTaggart the *A* series is crucial if time is to be real. Because of this central role, in § 4.3 I look closely at the components of the *A* series. In § 4.3.1 I find that although the concept of past has important social and psychological consequences, as a separate phase of change it shows itself inaccessible to coherent description or proof. In § 4.3.2 I find that present, like past, fails any attempt at sufficient description. In § 4.3.3 I conclude that the fact that future cannot be known until it becomes present causes us to suspect that there is indeed any such thing as future, and that we have produced the term only as a means of explaining away the puzzling re-occurrence of 'more' present. From this, in § 4.4, I conclude that none of the phases of the *A* series is independently acceptable.

4.2 Our concept of time

It is part of the human condition to be concerned with the nature of time. We cannot imagine anything without placing it in or invoking time. Our commonly held view of time is drawn from what Le Poidevin and MacBeath call the three central concerns: our belief that change is going on, that changes are caused and that there are limits to the extent of change.[1]

In primitive societies metaphysical reflection was not as important as survival.[2] When they did reflect on the nature of the world, they believed they could only systematically investigate it, if they could somehow impose permanence on an apparently ever changing universe.[3] This deeply rooted belief in the inescapability of change has had a strong influence on our concept of an eternal existence beyond the world of appearance.

There are two main strands of the human concept of time. Its chronology and its nature. It is the nature of time that causes us to reflect on past, present and future and earlier and later. Most of us today have a view of time sensitive to a daily chronology; we can call this a *Q* series. The *Q* series is my term for the metric of time (as I consider it *Q*uestionable as a means of analyzing the nature of time), it is a method of timing and should not (as is often the case) be misconstrued as to do with the nature of time. The *Q* series may use the *A* and *B* series as reference points but its main purpose is to measure time's passage. We also share, by commonly understood language and meaning, a general view of pastness, futurity and presentness, an *A* series, and a concept of earlier and later, a *B* series. The contemporary Western view of time is a blend of the rigorous 'fact' of the serial passage of time in the *Q* series, and the various appreciations derived from our experience of a 'sense of time' in either the *A* or *B* series.

Time dwells in our common experience. Faced, from generation to generation, with the knowledge of our parent's decline and demise, we are forced to observe the ravages of change. As we build on an increasing personal history and anticipate a depleting personal future we sense the 'loss' of time. It is not surprising that the 'river of time' appeals strongly to our consciousness. Many things in life we can apparently control, but there seems no choice involved in tomorrow following today. It is little wonder that human folklore abounds with adages about time. Indeed, our concept of time itself is an unavoidable anthropomorphic projection of the human mind.[4] How much this projection reflects something that is truly temporal is usually beyond the scope of our everyday concerns.

This deeply rooted human concept has led to two groups of philosophical thought; those that believe in the reality of time and those that do not. Of those that believe in it there are some who think that it is a universal constancy, a 'river of time', which we move through or that flows past us. There are those who believe that time is real but is inseparable from space. Some hold it to be regular and some hold that its pace ebbs and flows. Some think that future and past coexist symmetrically and some that the coexistence is asymmetrical.

These views have been reflected in philosophy from the early Greeks to the modern day. Heraclitus thinks the world is the totality of events and not things. Aristotle believes that there is a basic temporal structure and comings-to-be. Plato believes that time is not a pre-existing skeleton upon which the universe is hung, it is part and parcel of the creation. Plotinus believes that the image of time is an erroneous representation of the unity of eternity. Newton holds that space and time exist independently of objects and events that are in space and time. Leibniz and Mach hold that there is nothing to space and time other than the spatial and temporal relations that objects and events hold to one another. Einstein believes that neither time nor space is independent but exist together as space-time that is relative to other properties of the universe.[5] There are various reasons why some deny the reality of time. Parmenides thinks that the ultimate physical reality is timeless, Archimedes that time is not intrinsic to reality, Hegel that time is incoherent. Even the most ardent in this group, however, find it difficult to admit that there is not something happening that seems like time. Admission of psychological existence demands something associated with a temporality of some sort; it is impossible to admit our existence without also admitting a passage of time for the admission. When we reflect on time we sense a deeply rooted problem; how can we reconcile our psychological experience with the actual nature of time itself? This can lead us to confusion if not despair. The difficulty can cause us to make arbitrary judgements or even, as we become increasingly bewildered, like St Augustine, ask for divine assistance.[6] Even if our everyday appreciation of time is all that time adds up to, still we suffer from fundamental confusion. Hume believes the problems of space and time, though perhaps products of our mind, are unsolvable.[7] In other words, even if we feel that we know what time is, it is unexplainable. McTaggart takes a much more sensible approach. He thinks that time is incompatible with its own terms and therefore does not exist. He believes that we sense something that seems like time, and that this is a real thing, but he holds that it is not temporal.

But is the 'real' nature of time so certain and yet so abstruse that our everyday understanding of it is just nonsense? Are we entitled to theorise about something that has such fundamental human importance and yet allow that the theories are not to be commonly understood or interpretable? Or is that itself nonsense? We do not after all demand an intimate understanding of the workings of machines to be able to operate them or appreciate their general utility. But can we apply this to time? Time is not after all (as far as we know) assembled by someone who could provide an explanation of its workings. The fact that we can live with and use machines with confidence somehow implies that we believe there to be an assembler (or at least designer) who originally understood how it works. No such claim can be made for time. We might be completely mistaken about its nature and though our interpretation of it more or less fits with our temporal experience this does not guarantee that our interpretation, in any true sense, fits its nature. As Ryle says about self-consciousness,[8] though we may know what we are about *with* time this does not entail our making any sense *about* time.

The philosophical view that time exists further can be divided into two main strands. That either time is absolute; that is something that exists in-itself quite apart from any other influence or determinant, or is relative; that is that it relies for its existence upon another part of the universal structure.[9]

The term 'absolute time' defines time that is independent of events. Newton describes it as, 'Absolute, true and mathematical time of itself, and from its own nature, flows equably without relation to anything external'.[10] Absolute time should not be confused with either 'universal time' or 'cosmic time'. Universal time is always relational. It refers to a universal acceptance of a time series so that everyone on earth agrees that they are experiencing the present moment simultaneously. Relativity theory makes such a universal frame of reference impossible. Cosmic time refers to all time that has elapsed or is to come in the existence of the cosmos and essentially is a *Q* series.

Absolute time has a nature not reliant upon anything else, but this seems suspicious. The definition ascribes to time the function of flowing in terms that are themselves part of the definition. If absolute time flows it includes, as part of itself, a series of events in time and this seems meaningless. In addition, for time to flow equably or uniformly implies an external 'controller' of that uniformity. In itself this is not unreasonable but it would mean that time did not flow 'without relation to anything external'. If this were not so, and there is no external reference against which that flow is measured, there can be no sense of either non-uniform or uniform flow. Either term is meaningless if measured only against itself.

Absolute time, for all its drawbacks, provides a fixed time measure. This is an attractive feature because effectively it regulates time which although intuitively deemed to be regular, appears to our experience as highly irregular. Newton draws his conclusions from what he considers empirical proof about motion (his famous bucket thought-experiment) and so rests it on scientific certainty. Newton's principle was at one time supported by Russell who argued that the position of an object is uniquely determined if it is given at a particular time. If on the other hand the position is not given, then there could be an infinite number of moments at which it could be in that position. Russell judged this to mean that time must be a sequence of absolute moments against which events could be correlated. He later abandoned this view.[11] Berkeley thinks that Newton's assumption, that the experiment would be the same in empty space as it is on earth, is invalid — other cosmic movements must be considered.[12] Mach claims that, to achieve any sort of correct result, the bucket would need to be at rest with the universe rotating around it.[13] Sciama points out, 'It would follow from Mach's principle that inertial frames are distinguished not by being absolutely unaccelerated, but rather by being unaccelerated relative to some suitably defined average of all the matter of the universe'.[14]

For absolute time to exist we have to accept that moments (as entities in themselves) form the fabric of time and that events are descriptions of what happens in relation to that fabric. But there are problems when we identify time with objective entities like physical moments, and consequently with either motion or space. Cleugh picks up the 'undesirability' of this trait in Dunne's comments on the difficulty of imagining 'travel through an empty and ineffectual continuum'.[15] To be handicapped in this way in our reflections on time is a serious drawback; it shackles us to a very limited view of time as a physical (absolute) entity. There are more possibilities for time away from this standard. For example, we could entertain the idea that an absolute series of moments is unnecessary; that events 'happen' just because they happen to 'happen'. That they correlate themselves because they coexist but do not need an entity, 'moment of absolute' time, to do this. On this view, we would speak of a moment because it happens and give it no additional qualities. Such a moment is not a temporal entity existent in its own right, it is part of the class of coexistent events themselves.

I think we cannot usefully get involved in any theory of time that places moments so highly; we derive time from events, not vice versa.[16] We can, however, be misled if we believe that time relies upon events. It may be that time cannot exist without events but events themselves must be defined in such a way that they are not inherently physical events. If events are part of a physical world and time relies upon them, then time is necessarily reliant upon the physical world. As such, an event is usually said to exist at certain moments or instants as opposed to certain other moments or instants. Russell (in his later philosophical period) defines events as 'each occupying a finite continuous portion of space-time. It is assumed that two events can overlap, and that no event recurs ... Whatever is earlier or later than something else I shall call an "event"'.[17] This is much too limiting and we have more latitude (assuming that increased latitude favours us in arriving at a sound conclusion) with McTaggart, who allows for any kind of existential entity to form an event as long as that entity occupies a position in time.[18]

Even if the universe consists in 'empty space' this does not rule out the idea of absolute time. Absolute time still could 'tick away' in a physically empty universe as in a full one; though it would be difficult to argue that the innate existence of empty space formed an event. So although, in an 'empty' universe, absolute time could form a background temporal mechanism, it could not be applied to events, without which there could be no change (except in absolute time itself) and consequently no 'time' as we experience it. In order to accept absolute time, we must (like Kant[19]) assume that it is given *a priori*, so that although in our experience, nothing can be apparent without time, we could remove all appearance and be still left with time itself.

If we accept absolute time we have to be able to live with the idea that the universe has one 'time code'. It may be serial (perhaps as a *Q* series), it may only be a present or it may be a present and a future or a present and a past. Whatever it is, it will have a uniform application throughout the universe and such a strict metric, though born out by clocks, does not sit comfortably with our temporal experience.

'Relationism' is any theory of time that allows that events are more fundamental than moments. Here, time has some degree of dependence upon what appears to be happening in time. As Lucretius says, 'It must not be claimed that anyone can sense time by itself apart from the movement of things or their restful immobility'.[20]

Leibniz's 'reductionist' view of space and time typifies what is central to relationism.[21] On his view, all true propositions are analytic (that is the content of the proposition is already contained in the thought) and reality can be reduced to independent perceiving units he calls monads. We can take Leibniz's 'monads' here to be either physical or mental entities. Each monad operates during its existence according to the pre-established, God-given harmony.[22] He distinguishes between things in time, events and moments or instants (which he classifies as temporal items). He believes that time is a succession of events and always can be understood in terms of events. Time consists of a first event after which all other events have followed in strict sequence. Any investigation into the nature of time thus becomes empirical. It does not involve mysterious and unsolvable questions about the nature of such things as instants. The consequences of this reductionism are that both change and temporal connectedness are essential — different events must keep following previously different events. There can be no temporal vacuum in such a universe. Leibniz's rationalist view is not a response to Newton's bucket thought-experiment and is based on his principles of a Sufficient Reason, of the Identity of Indiscernibles and of Pre-established Harmony.

In this system, if we suppose that empty space is something absolutely uniform (no part of it differing absolutely from any other part), then God could have had no reason for placing the existing configuration of bodies in space any other way around than they are. If, on the other hand, we suppose that the relation of bodies one to another constitutes space, then placing them the other way around would not make any difference. Our error here lies in our incorrectly attributing a quality to space in-itself. Leibniz thinks that the case is the same for time. He thinks that although God may have created the universe a year earlier than he in fact did, it cannot be known because time is not something separate from things.[23]

Leibniz's monads somehow mirror the state of the rest of the universe, which is made up entirely of monads. Because of the mutuality of this mirroring, each monad must be in a corresponding state to all other monads at any given instant. This mutual accord is perfectly set by the condition of pre-established harmony and because of the perfection of God, the harmony will continue. For Leibniz, neither space nor time can exist independent of the harmonious accord of the monads other than as ideas in God's mind. Time can only be a succession of events. Although the monads exist in a universal 'time', that time is set by them all being in the same condition at the same moment. There is therefore no reference to any absolute time that is external to themselves.

In some ways, McTaggart's atemporal world is similar to Leibniz's relational world. Like Leibniz's monads, McTaggart's 'selves' arise from an originally harmonious oneness. They proceed through a non-temporal succession, the *C* series, moving all the time towards a full perception of all other selves. When this end-stage of complete perception is attained, the succession ends and the universe remains in a static and timeless harmony.[24]

Unlike monads, however, selves do not share a universal time (or more correctly a universal simultaneity); there is no successive simultaneity in the *C*-world where no self shares a common point with any other self.

The rationalist notion of relationism has dovetailed into modern scientific thinking which now has taken on much of the responsibility for answering questions about temporality. The identification of the 'real' with the outcomes of modern physical theory has reduced the scope of metaphysics and perversely led to a marginalization of rationalism. Reichenbach encourages us to throw out the belief that idealism is the only true epistemological position when thinking of time. He says that mathematics can provide the conceptual structure, which is the ideal, that physics can then coordinate into reality. His idea is based on the assumption that time can only be viewed in relation to space.[25] Gödel suggests a modification to this view. He adjusts the place of metaphysics, lining up the ideal of the mathematical structure more closely to the non-material idealist view of which McTaggart is the epitome.[26] The general acceptance of the terms of theoretical science has led to an almost unassailable belief in the objects of those terms. The 'building block' idea of the ultimate constituents of the material world, started in the early nineteenth century with Dalton's atomic theory,[27] has led to a profusion of particle theories, each apparently more fundamental than its predecessors;[28] each re-analyzing the structure of its 'indivisible' predecessor. These days, it is the minority who think, as Keeling does, that, 'We must not permit a mathematical device to settle a problem which is metaphysical'.[29]

The idea of space-time has been very influential on twentieth century thinking. Much of the theoretical work in this field is attributable to Einstein and is explained in his Special and General Theories of Relativity.[30] However, as Newton-Smith points out, it is not at all clear cut that the Special Theory of Relativity counts heavily, if not decisively, in favour of the Leibnizian-relational (reductionist) conception of space and time over the Newtonian-absolutist (Platonist) conception.[31]

Space-time theory relates space and time as one class and then relates this class to all other foci in the universe. By making these relations, time, as space-time, appears both un-simultaneous and irregular. The assumption of the Special Theory is that the velocity of light is independent of the motion of the observer and is distributed uniformly in straight lines in Euclidean space.[32] An observer associated with an inertial frame of reference, equipped with standard clocks and measuring rods, will record the local speed of light as the universal constant *c*. The velocity of light is considered the highest speed that information can be transmitted between existents. Higher speeds can be considered, but the velocity of light remains the maximum speed for the transmission of the information. In the absence of gravity, each observer in an inertial frame of reference will record the velocity of light non-locally as the universal constant *c*. The General Theory of Relativity considers the influence of gravitational forces. It shows that in the presence of gravitational forces the velocity of light is no longer uniform, although local velocity, when measured by standard clocks and measuring rods, is always the same.

The Special Theory flies squarely in the face of common sense. For example, if we travel in an airplane at 500 miles per hour relative to the ground, and a bullet comes past us at 800 miles per hour relative to the ground, we imagine that it will pass us at 300 miles per hour relative to our own situation. But, by the Special Theory of Relativity, if we are travelling at .5 the speed of light and a bullet comes past us at .8 the speed of light, its speed is approximately .8 the speed of light relative to both us and the ground. The superimposition of two velocities of less than the speed of light will always 'add up' to less than the speed of light, .8 light speed plus .5 speed of light equals less than the speed of light. In consequence, time dilates, this means that the faster a body's velocity, the 'slower' its own time passes. In other words, if you have a clock flying through space at .9 the speed of light, the hands on its face will travel much slower than the hands on a clock resting on earth. Though of course the earth itself is moving relative to other bodies in the universe and its own velocity is only known in a relative sense.

Space-time of the Special Theory exists in Newtonian space and so bears out the absolutist model. Time in the General Theory does not. In recent years, there has been more interest (both scientific and philosophical) in how time can be understood in relation to the General Theory. Having lost the concept of simultaneity with the Special Theory, we lose the concept of *any* temporal independence in the General Theory. Its 'fabric' therefore offers greater, more dynamic epistemological opportunities than a three-dimensional space plus a one-dimensional time. It offers a four-dimensional framework within which there are opportunities for two-dimensional time,[33] time of a non-serial nature and time travel.[34] Many say that it is inappropriate to deal with the nature of time without using the basic, four-dimensional framework of General Relativity. I think that in some ways, this framework equates with the idealist view (see § 5.4), though we cannot, strictly speaking, deny time if we accept this, because it does not exist as a thing-in-itself in General Theory.

When it comes to thinking about time, we are not confined by the parameters of either absolutist or relational theories; we have a third choice

— the non-existence of time. Admittedly, even if we deny time we can still analyze the appearance of time either in relational or absolute terms. For example, though Leibniz holds a relational view of both space and time, ultimately he thinks them unreal. But if we deny time and are not so interested in analyzing its appearance (which applies more closely to McTaggart) then relationism and absolutism are of limited interest; doing this would be like analyzing the symptoms of a disease and not the disease itself in the hope of finding a cure. Both the absolutist and relational views of time accept that there is a 'real' time (even if only as appearance) active in the universe that can be explained by use of the appropriate terms. In its barest form, the non-existence of time is the view that time is a 'figment of our imagination'; that it is just a method of interpreting the world so that we can order our ever-changing place in it sensibly. This is not, however, the same as, for example, using the terms 'happy' or 'sad' to explain our mental states so that we can share a common understanding of those states. If there is no consciousness to appreciate 'happiness' it is reasonable to assume that there is no 'happiness' in the universe; a universe without consciousness would not seem to suffer from having no happiness. Indeed, happiness would not seem essential to its progress. On the other hand, if time is just an expression of human consciousness we have more difficulties than in the case of happiness. If time is just a conscious figment then without consciousness we must be assuming a universe without change. There could be a universe without change that exists at t^1, t^2, t^3, and so on, but it would be the same universe at each successive moment and each successive moment (which is a time series) would have in some way to be separate from the unchanging universe. If this were the case we then are dealing with two 'sub-universes' which combined together form a universe that includes time. Though such a 'time sequence' would be a way of denoting a succession of similarity it conflicts with Leibniz's principle of the Identity of Indiscernibles and a good deal with common sense.

Granted, some reports of no change make good sense. We could say that 'there has been no change in the patient's condition' but this is a change of a different sort. There may have been 'no change' in the patient's condition but all this means is that we can judge the general nature of events surrounding the condition of the patient yesterday to be the same today. But we are only using this judgement for ease or efficiency and only in a general everyday sense. In fact, the patient is a day older, has had some degree of worsening or lessening of his condition (though apparently nothing that would cause any alarm or relief), has different contents in his alimentary system and has telephoned his mother to wish her a happy birthday. The notion of the non-existence of time cannot mean the denial of a framework in which change, whether perceived or not, occurs. What we mean by the unreality of time is that what we generally think of as time (in the *A* and *B* series) is erroneous and that whatever the universe is changing 'in' it is not changing in time. In this way we can accept the fundamental aspect of universal change but deny the temporal connotation that we find both convenient and plausible.

But is this really a third alternative? Can we reasonably shun common sense so easily? When dealing with the nature of time common sense provides a useful safety net; if we are faced with a problem that is as confusing as time then we tend to fall back into it. Common sense in itself cannot claim to arbitrate successfully on subjects which confuse us but it is a psychological standpoint that should be reckoned with. Sometimes common sense is used incorrectly; it may be simply an apology for action or a justification for views or actions which have not properly been worked out. On the other hand, appealing to common sense in the face of seemingly insurmountable confusion can be meaningful. In such cases we are drawing upon something deeply imbedded in our subconscious and that this gives us some sense of 'rightness' should not be ignored. If we have an intuitive awareness of something which we find difficult to articulate (for example, the nature of time) then common sense is always worth taking into account. In addition, a theory is weakened if it has no common sense appeal. Price provides a good definition of common sense as, 'a body of very general principles commonly accepted by ordinary non-philosophical men in the ordinary affairs of life'.[35] It is an important consideration. Its principles are really philosophical, metaphysical, epistemological or ethical, but applied without any awareness of this. These principles together form what is usually considered the common sense view of the world. The common sense thinker, however, does not deduce principles from anything else but accepts them in their own right without involving any explicit acknowledgement. Indeed, we neither increase nor decrease the common sense acceptance of them if the principles are formulated. If we deny a principle such as 'time exists' the common sense thinker is liable to protest. This protestation would be based on a deep-seated belief that time is real and to say otherwise is nonsense. In saying 'It's common sense', is to draw upon a large body of agreement.

The common sense thinker, though he may understand that a philosophical principle is involved, does not become involved in formulating the principles used and he can often be a bad judge of such formulations. This makes him prone to questionable philosophical arguments. But the philosophical weight of common sense should not be ignored. It could be said that the foundations of logic rest on such unformulated acceptances.

So where do we find this common sense thinker? We could stop someone in the street and ask 'Is the denial of the reality of time a reasonable view to hold?' He may have been watching a television programme on relativity and be content to accept that 'It's all relative isn't it?' By saying this he would be genuinely dismissive of the question. Alternatively we can look into ourselves and appeal to our own intuitive common sense. But to cross-examine ourselves in such a way will always expose the philosophical bias of the examiner. We will tend to get the results that we want. Another way of investigating common sense is to examine ordinary language, but this also presents problems. We often use several words to stand for the same object or thought. There is the danger of hypostatizing abstractions. We need to consider contexts as well as knowing what is referred to in the first place. Yet, on balance, this may have more 'objective' merit than direct introspection that does not involve the 'checking' system inherent to the analysis of language. But the analysis of language only presents us with the particular applications of the principles. Our understanding of the principles themselves is drawn from a long heritage of philosophical endeavour. This has led us to firmly established common understanding of major concepts such as cause and effect, relation, substance and so on. Though even here, still we have doubts.

The common sense thinker is also liable to inconsistency. Time seems to be an unalterable process in the *B* series, always passing from earlier to later. But this does not prevent the common sense thinker from entertaining the idea of travelling back in time and making corrections to his past that will affect his present. It may be logically possible for me to travel into the past and make major alterations, for example killing my parents so that I never get born. But even this could not alter the fact that I was there killing my parents and so, whether or not in the past I was unable to be born, this could not alter the fact that still I existed.

Nevertheless, though it is the common ground of understanding, common sense is not a primitive sense. It includes much breadth and depth of understanding and it has authority. Bergson thinks the common sense belief in measurable time is unreasonable; for him, time is simply a manifestation of our imagination.[36] Psychology leads us to think that many common sense views are products of our evolution driven by biological needs. If the biological needs had been different, our common sense views would have been different. There is, however, nothing necessarily 'savage' about common sense. If an argument appeals to common sense it has a seal of approval with lineage to pre-history. Common sense is not philosophy but philosophy should respect what it contains. That the common sense thinker does not 'know' philosophy does not make his opinions on philosophical principles invalid. There is what Price calls a *consensus gentium*,[37] a worldwide agreement in favour of common sense that goes beyond a description of its contents. Assumptions which have evolved throughout our history cannot be set aside as capricious; they deserve appropriate and serious attention. On this basis alone we should not be troubled by using the statement 'it's common sense'. It is a reasonable principle to invoke when dealing with something apparently so often at odds with common sense as time.

4.3 The components of the A series

According to McTaggart the *A* series is crucial if time is to be real. Whilst we maintain a belief in the *A* series, we will always be drawn, by common sense if by nothing else, into a belief that it is real. In this section I look closely at the components of the *A* series in order to show that no part of it stands up to close analysis. If, as I believe, each *A*-phase is transparently erroneous, we can combine this more 'common sense' denial with McTaggart's argument for time's inherently contradictory nature. Attacking time on two fronts like this 'doubles' our reasons for doubting its reality.

The term '*A* series' describes the direction of time in a one dimensional transitory series. That time is directional in itself, or somehow has relative directionality to other things, is not universally accepted. For example, Reichenbach links temporal order to entropic theory, bringing together a space-time topology and a causal world structure.[38] Though we may, like Sklar, consider this 'an epistemologically generated type of "philosophical" reduction',[39] such a view shows us that we do not necessarily have to find the terms of time in abstract qualities that seem to resemble time itself. Nevertheless, we tend to believe in the *A* series so strongly that we use science to affirm it as well as try to fit it into scientific structure. This is compounded by those that think that the *A*-fabric is so real it can be subjected to the sort of scrutiny that others think can only realistically be applied to the workings of clocks. The notion of backward causation exemplifies this sort of thinking inasmuch as it assumes the reality of the *A*-fabric so strongly that it uses it as material for alternative world views. This then leads to progressive argument not about the reality (or otherwise) of the *A* series but about the ways in which it can exist. For example, Earman takes a strong *A*-

view,[40] Dummett develops Earman's arguments applying them to Newcomb's paradox and argues in support of backward causation,[41] Riggs supports this view,[42] Mellor argues forcibly against backwards causation.[43] The *A* series sits at the centre of this; acknowledged as if in some way absolute and available only for inspection and interpretation. It seems that it takes more than McTaggart exposing its contradictory nature to shake it from the roots of thinking about time. It is for this reason that I will now go on to show that not only is the *A* series self contradictory but each of its phases, when analysed, dissolve into confusion.

Although there is much philosophical disagreement on the subject, to common sense our experience of the apparent direction of the *A* series is its most obvious and inescapable property. If we accept that time has a direction, are we to assume that there is only one directional series? Indeed, there may be more than one *A* series, either a different layer or a two-dimensional version. Why then do we naturally tend to support the idea of only one time series? After all, if there is one time that is so epistemologically certain, there is no reason to doubt more than one time. This is important for the reality of the *A* series because if there are several *A* series operating at the same time then time's passage, the nature of change, the direction of time and in the end, the concept of a discrete regulating *A* series itself breaks down. Quinton sets out to show that whereas we could make sense of lives in more than one space, we cannot make sense of lives in more than one time.[44] Swinburne originally thought that he could show a multi-temporal meaning in his myth about the Okku and Bokku,[45] though he later decided that such a scheme entails a denial of logically necessary truth.[46] Support for Swinburne's original view is found in Hollis.[47] There is nothing contradictory about there being any number of times-in-themselves, but it is difficult to imagine an experience of more than one time. To experience this would make our lives impossible to order or understand.

It could be that time itself is not so closely related to the experience of time. Our thoughts, stimulated by our experience of time, may indeed be ways of thinking about how time actually is. But, at the same time, time-in-itself may have a nature that cannot be reduced to our experience of the relations of things or events. After all, time-in-itself still would determine those relations whether or not they reflected its actual nature. This Kantian (in this respect absolutist) view is developed by Swinburne who highlights the idea of time (and space) having properties, independent of things that it contains.[48]

4.3.1 Past

Common sense tells us that things that happen in the present are 'real'. We think we can influence the present by controlling events that occur in it. We cannot influence the past that is 'dead and gone' and we cannot influence the future because we 'do not know what it contains'. We think, perhaps more rightly, that we can do things now that may affect what happens in the future. While we are, however, in the present we cannot expect to alter things by 'digging into the future' any more than we could by 'digging into the past'. Common sense also tells us that we exist and act in the present, although it acknowledges the strange nature of 'nowness' and how it could be in the past by the time we can come to know it via our senses.

Indeed, the past may be as real as the present. One idea is that what happens to the present by becoming past is that present moments become consolidated in the ever-growing reality of the past. As Broad says, 'fresh slices of existence have been added to the total history of the world'.[49] In this sense nothing has happened to the present by becoming past. The only change that has occurred is in the past and that change only amounts to a new 'slice' of present being added to it. In this way the past is as real as the present. Such a past is always increasing as new presentness keeps moving into it. Because we are aware of this increasing nature of real past existence we derive an idea of temporal order. Also, as the present moves into the past, we are aware of the increase in the sum total of the past (and reality). A by-product of this is an awareness of *B*-relations. We are made aware that what is now in the past was earlier than what is now in the present. In this way we arrive at the same conclusion as McTaggart, that our fundamental sense of time is reliant upon the *B* series. We therefore accept the reality of time and place all that exists in one part of a diminished *A* series, the past. The whole thing is held together with terms, which McTaggart says, are incompatible. Though Broad only allows the two terms past and present, McTaggart thinks two terms are just as incompatible as three and therefore just as contradictory.[50]

Both views accept the classification of past and present, though one denies future. Both agree that there is a movement 'from' present 'to' past, though McTaggart believes this movement initiates a vicious infinite regress because of the implicit self-referring terms. Both agree that the past changes, Broad because past is the only reality and McTaggart because of the continuous action of *Cambridge change* in events.

In *A*-terms there is nothing that compels us to believe in the future. The *A* series does not suffer from its loss. A movement from present to past is equally as contradictory as a movement from future to present to past. On the other hand, we cannot reduce the *A* series only to past and be left with something that makes sense. Because we know that we do not exist in a static past, any reduction like this leaves us with something that is clearly even more erroneous than the *A* series containing past, present and future. At least the full *A* series seems to reflect in some way what is going on.

Common sense always denies the idea of the past changing. If there is one thing as inaccessible as the past then, on the face of it, it can only be the future. But the future is not quite so inaccessible as the past. If we wait, the future will enter the present. The past however, will only go deeper into the past. It is open to the memory but inaccessible to further experience. To common sense the past is always closed.

Whether time is cyclic or a system of eternal recurrence, the past is irrevocable. It would be startling if the past began happening again. Though in some respects if it was the 'real' past (as opposed to a 'mark II past') which was happening again, we should be none the wiser. This could be happening continually without us knowing. If it was, it would confuse (by combining) our understanding of 'continuousness' and 'repetition'. Our experience would be continuous and therefore moving always into the future, whereas events once experienced in a previous time, would be repeated. An alternative way of experiencing the past would be to jump from one universal point to another (as Gödel suggests[51]). As we dodge from one universe (and point in time) to another, the occurrence of pastness at different points in the structure of the continuum makes the experience of the past possible and accessible. This type of movement in different spatio-temporal dimensions features high in contemporary science fiction (for example, *Star Trek* and *Quantum Leap*) to say nothing of the theories of 'worm hole physicists' and popularisers of scientific cosmologies.[52] However, it remains that our experience does not go in reverse; we may imagine being in a part of the past but we always experience it in a conventional *A* series way.

Also, we may think that all existence is contained in a past that grows, and therefore changes, continually as each new present moment slips into it. Such a past is full of all previous presents and as such must be the total of existence, except perhaps for the one and only present, present moment. In this way, past is an unavoidable certainty, the fabric of which, in volume, and therefore possible importance, far outweighs the ever-moving fragment of present.

Though McTaggart thinks that everything eventually ends up in the past, the true nature of that past is not our perceived temporal past in the *A* series. For him, the past 'absorbs' all future events that have become present and then past. The past is therefore always changing. He accepts that this view relies on the reality of the *A* series, but thinks that it is independent of the view that all change depends upon the *A* series. He says that at each moment, every past event becomes further in the past than it was before and though this is an outcome of the *A* series, it is 'independent of the truth of our view that all change depends exclusively on the *A* series'.[53] On this basis, the changing past can be deduced directly from the *A* series, quite separate from the idea that if the past is changing, then it must be because of the *A* series just because all change derives from the *A* series. This conflicts with common sense which generally combines the view that past, present and future are real, but that the past does not change.

When we think of change in our daily lives, we think of being involved in it, of being able to alter things according to our own volition. There is no opportunity for us to be involved in this way in the past. It would be an unusual view that proposed that we could 'claim' and actually 'use' past events in the present. We use the past (that we remember) as information in the present, but nobody really believes that we actually bring past events 'back from the dead' to be currently usable in the present. Obviously, this would not be the case if the *A* series was cyclic, for example, Dummett's view of backward causation invokes a cyclic interchange between past and future.[54] But the terms of the *A* series are separate and serial and do not allow for any such cyclic interchange. If this were so, no longer would we be talking of an *A* series which is based on experience but on an *A* series of which we have no experience. Still more improbable is for us to somehow leave our presentness behind and go 'into' the past and there do only 'past things'. It is a common fantasy to think that we can go into the past but this usually takes the form of dealing presently with things that we (or we once thought were) past. If we are to know that we are in such a past then we must have some experience of 'presentness' with us. The demonstrative 'I' who is experiencing the past is the 'I' contemporary to my experience of myself. Yourgrau identifies this from Aristotle's statement, 'in the case of living things, their being is to live'[55] as 'Aristotle's reasonable observation, which is itself a straightforward deliverance of the intuitive conception of time and of living-in-time (and what other kind of living is there for those of us who are not gods?)'.[56] This is apparent in Descartes' *cogito, ergo sum* and St Augustine's *si fallor, sum* ('if I err, I am').[57] Though St Augustine's 'I' is more McTaggartian in that it is spiritual, whereas the function of Descartes' 'I' is to seek out reality of the external world, they both represent the notion that though I can think I do not have a body, I cannot think that I am not. I can only think that I am if I am experiencing myself in the present so any idea of doing 'past things' in the past is nonsense unless it is doing things in *my* present which simply appears to be contemporaneously placed in what I once thought was the past.

The idea of going back into the past by literally going backwards in time is different. This can come from mathematics (Gödel), from philosophical analysis (the sort of structure envisaged by Earman) or from science fiction, for example in Wells' *The Time Machine* or *Dr Who*.[58] In such a world we would start going backwards in the *A* series. Past events would move into the present and pass into the future. Everything that had happened in, say the last ten years, would happen again, but in reverse. There would be no question over whether it was the second time that the events of the last ten years had happened. They would definitely be happening, but this time they would be happening in reverse. After ten years of reverse *A* series the backward process could stop. Things would get back to normal and the *A* series would again start passing from future to present to past. Because, to arrive at this point (ten years before I had started) everything had gone in reverse, all the memories and experiences of the last ten years that I had originally (in the first positive *A* series) would have been eradicated. When the negative *A* series again went into positive mode, I would be literally living the time over again. Every experience of the present in this new positive *A* series, would be entirely new. It would not be a mark II 'now' or a repeat. It may be different from the 'first' now, because of what I did 'again' (in the 'second' past) ten years ago, but because the first now has been entirely eradicated, such a difference is meaningless. There is no first 'now' against which to compare the second 'now'. In this way there is no limit to the number of times and the number of changes that could be made to the past. Indeed, if the negative *A* series happened instantaneously we would never be aware of it happening at all and this would be happening continually.

The past can also change when incomplete information in the past changes into more complete information as more past is added. Take, for example, a tune made up of notes. The first one, by itself, means little as far as the tune to come is concerned. The hearer may recognise the note and extrapolate the tune from it. As the subsequent notes of the tune occur, the tune itself is revealed. In this way the complete tune is only known when all the notes have become past and take on their new form as part of the tune. The example is to be found in the description of this given by Loizou, he calls it the Principle of Emerging Identity.[59] However, though the tune is only known when all the notes have been lodged in the past, the knowing of the tune and the completeness of all the notes of the tune, occur at a time that is present and simultaneous to the knowing. This 'knowing' of the tune then becomes part of the past in just the same way as all the individual notes are past. This sort of distinction, between the individual notes and the tune they all make, has limited use. Our present comprehension of a whole tune is utilizing a memory of the past. It involves the preceding notes and not elements that are themselves part of the past.

The notion of the past changing seems paradoxical. However, that the past does change is within the truism that there is at time t^2 more past than at an earlier time t^1. But this does not alter our view that we cannot change something in our past that we regret (by making amends or undertaking some other corrective action in our future present). If a more natural quantitative model is taken we are again misled. If 'the past' means particular events, now past, they would need to exist objectively, somehow separate from the flow of time, to be recoverable. If such objective events could be recovered, although the experience of them would be a mark II experience, they are imaginably experienceable. If the past is the totality of such events then any further experience of them would mean an experience of all that has already 'gone' into the past. Such an experience, either cyclic or eternally recurring, again would go against our only experience of the asymmetrical *A* series.

The flow of time in the *A* series seems inescapable when faced with our common sense and overwhelming empirical evidence. But the past *does* change. At the same time as we pronounce it 'past and therefore unchangeable' we realise that more present keeps flowing into it. As this happens it alters by more facts being added to it. But this is to do with the way that we classify past. We are looking at two different types of past. One, that is the class of all things that were once present and before that, future. In this sense it is a class of moments in time in their final stage in the *A* series. The members of this class are past and, because of the strict serial nature of the *A* series, they can never again become present or future. The other, is the past as a class of all past events. This has an apparently infinite capacity to absorb as many new past events as there may be available. In a finite time series the past will come to a point where it will no longer grow. Either because the supply of used present moments dries up or the past simply becomes full and can take no more. If the past can take no more, its inability to absorb more pastness would occur prior to the ability of the future or the present to produce more past. In an infinite series there is no reason the past cannot remain infinitely available for the inclusion of further past presents. The changing nature of the past, in this respect, is like adding drops of water to the ocean but never changing the idea of the thing that is the ocean. But a past like this is always changing even though the moments added to it, to make it change, can never be returned to the present. No matter how many more moments are added still it will be always the class of past moments that we call the past.

The past can be seen as encroaching upon the present. In the *A* series, the future will, in due course, become present and then past. If the past is growing and the present does not grow then the past may be encroaching upon the present. This past would have to grow in a fixed dimension with a determined limit to the distant past. As it grew, more recent past moments would begin to get squeezed up more closely against the present. This would result in either the present going more quickly, because the past is 'squeezed' up tighter to the past side of the present, or the movement from future to past becoming more rapid.

If the *A* series is finite and the movement from future to past is like the movement of sand down an hourglass, the past will, in the end, simply fill and events will stop happening. The fact that the pile of sand at the bottom of the glass steadily piles up, and therefore encroached on the present, will not affect the passage of sand through the neck unless there is a difference between the volume of the future and the capacity of the past. If the capacity of the past cannot hold all the future events (that run through the present and into the past), then when the past is filled it will prevent the flow of future events through the present. At this point the *A* series would stop. It would leave a stock of future events available but prevented from occurring because of the choked present. Because the past has stopped, time can no longer be seen as made up of past which is the preceding 'present'. The only place where there would be any time (in the sense of temporal movement) would be in the future as future events try (but fail) to approach the present. This would be an unusual account of time where everything was only ever future with no potential for being anything else.

Another alternative is that the future actually changes into past. Imagine the *A* series to be finite, the capacity of the future and past to be equal and the moments of transition from future to past to be represented by future changing into past via a metamorphosis in the present. Like a box of sand changing colour from red (future) to green (past) with a quick flash of pink (present) as it changes. In this process the past will grow as the total amount of sand becomes increasingly green. As it grows it will encroach upon the future and ultimately squeeze out all possibilities of future events occurring. The last grain will change from the red of the future, flash pink and then become permanently green. At this point all of the *A* series becomes past and can no longer be called the *A* series. McTaggart's idea of the *C* series reaching a final stage is not unlike this.

If the *A* series is infinite then there will be an infinite supply of red in the future, just as there will be an infinite possibility of green in the past. Here we could assume that the amount of past will continue to become more than it was. However, it would remain insignificant relative to the infinite future.

But all these ideas expose the problem of identifying the past as some separate entity that receives 'used' presents while thinking it has a hand in the process of this consumption. If it is a separate phase that receives moments that are 'used' from the present, it is unlikely that it is involved in the process of change from present to past. It seems unlikely that the *A* series somehow alters the present by making present moments into past moments. This sort of past is like a dustbin and dustbins do not usually satisfy their own appetite for rubbish by creating it themselves. They only become the rubbish in another dustbin when their own useful lives end. This container view of the past gives us the idea that the past is a more substantial entity than the other two classes in the *A* series and as there is nothing to justify this we could be in error in thinking it.

We commonly think of time in respect of our experience of it (or how we imagine we could experience it), but this view of the past hypostatizes it as a thing-in-itself. We imagine the future as say 'possibly nothing at all', the present as 'so fleeting it may not exist' but the past as 'absolutely certain and known'. This is notwithstanding that it is made up of the 'nothing at all' of the future and the 'all so fleeting' of the present only to become the 'totally inaccessible' of the past. This hypostatization can entrap the common sense thinker and the philosopher alike. Common sense says we are 'in' the present, 'look forward' to the future and that the past is 'what is gone', as though these three aspects of time are both distinct and real. The philosopher can do the same by erroneously recognizing the past as irrefutably real and thus become engaged in sub-dividing it into other real things. For example, whereas Loizou talks about the 'emerging identity' (changeable) aspect of the past[60] Gale talks about the tense-logical (unchangeable) aspect of the past. Both make the basic assumption of the reality of the past. Interestingly, tense logic was first put forward in a coherent form by Findlay[61] who charges McTaggart with using conditions of a tenseless language and imposing them upon a tensed one. We can therefore ascribe the foundations of tense-logic to McTaggart. As Prior says, 'there is much tense-logic in McTaggart, though Findlay was the first to see it as such'.[62]

If the past is the only real phase in time (and therefore deserves the hypostatization), then future and present, or just present, are part of the preparation events must undertake to become past and therefore real. This way if thinking offers an acceptable 'place' for all time as the recipient of everything that has happened — a sort of temporal mother's arms. But it implies either a great deal of error in the perception of future (and to a lesser extent present) and nothing like the same degree of misperception of the past even though the past is inaccessible to perception except by the recall mechanism of memory. To accept it we would have to believe that there are some strange conversion processes going on as events move from the future to the past. In each transition the temporal nature becomes something that is somehow more real. If we accept the *A* series, then all past was future and present at an earlier time and cannot be credited with an existence that the other two are not. Although McTaggart considers us in error in our perception of time, he does not think we are in error about one part of the *A* series but not another. By throwing out time altogether he does not therefore have to tackle the improbable transformation of parts of time from unreal to real.

A further problem about the 'extent' of the past is that things become further and further past.[63] When we think of past in the *A* series our common sense tells us that there is a difference between recent past and distant past. But if past is the class of events that were once future, then present and now past, they are either past or not. If they are past then they should be part of an homogenous past that does not discriminate between relative states of pastness. If they are not, then they are either future or present. To discriminate between recent past and distant past is to extend the seriality of the *A* series. The past itself becomes a temporal plurality with different points at different times in its own existence. The river of time takes control and produces a self-contained history of pastness. A simple analogy exposes such thinking as inappropriate. If sand grains are in a heap then as far as the description of the heap is concerned it is irrelevant where each particular grain is. All we need to know is that it is located somewhere in, and therefore forms part of, the heap. If the heap is only known to us as a heap that is inaccessible, the idea of locating different grains of sand is pointless. To imagine we can pick out 'grains' of pastness from the inaccessible 'heap' of pastness seems improbable.

There is also the problem of how our perception of past leads us to hold a view of past that necessarily fits in with our perceptions. After all, everything is known to us by some form of perception. Time offers no diversity of evidence that together forms a coherent pattern for assuming that something in time is truly what we think of it as. On the other hand, nothing seems more certain than that everything we perceive is in some way in time.

Thus, we are faced with deciding on the basis of phenomenology (which has intuitive appeal) or intuition (which has some phenomenological coherence).

It is easy to imagine that the idea of the past is nothing more than a feature of human psychology based on an ability to 'remember' events that have occurred. Without memory it would be impossible to have a coherent *A* series view at all. Generally we catalogue events as conveniently as possible. The 'oldest' memories are placed furthest back in the catalogue and the 'newest' memories placed at the front, nearest to the present. It is then convenient to view the present as a moving pivot between past and future, the movement being 'towards' the future and 'away from' the past. In this way our direct experience is always present. Our intuition of this process is that the past gradually increases to hold more experienced events. The future, either open or closed, finite or infinite, provides the supply of new events to be experienced. We often question the reliability of our memory. If memory is a big factor in our intuitive belief in time we should be suspicious of that intuitive belief.

In addition, much of how we view past and how we allow it to settle in our minds, is a product of a social view of pastness contemporary with our own existence. Past becomes an entity to be reviewed as a form of history in one light or another. These days, we tend to review the past as a series of events so strictly catalogued that it is held to be indisputably correct. What is not known or may be in error, will or can be known, if sufficiently investigated. This contemporary view looks at a future, much of which is modelled, and projected, by statistics from the past. For example, the weather can be foretold by receiving enough data from the past or we have an idea when we shall, statistically, die. Such future assumptions revolve around a *Q* series so important, that anyone who proposed time not to be real in a busy Western city, at a quarter to nine in the morning, would find few subscribers. Ancient views of the history of past are in sharp contrast to our contemporary view. Thucydides was not so much interested in the nature of time itself but invented what was then a revolutionary dating system for events in history.[64] For Homer past and future are intermixed and distended.[65] The contemporary scientific momentum which accentuates the importance of the *Q* series should not be seen as empirical evidence of time's reality.

The past, if it is to be real, must be more than just memory. If it is not then it is only memory. It is difficult to imagine that without memory we would bother filling up the *A* series with a concept of pastness that we could never understand. Without memory we would be perfectly happy to exist in an *A* series made up of only future and present. In fact it would be worse; it is unlikely that we would have any concept of future without a memory of past.

Without consciousness of past we would be doomed to a continual fleeting present that could have no meaning to us whatever. Though the concept of past has important social and psychological consequences, as a separate phase of change it shows itself inaccessible to coherent description or proof.

4.3.2 Present

Russell's views have led to a determined *B* series view that represents the alternative to McTaggart's notion of transitory time. McTaggart's view, though ultimately contrary to common sense (in that it denies time), has its roots more firmly in common sense than does Russell's that fails to reflect our psychological experience of the world. Common sense, when face with the confusing nature of time, allows for the idea that 'time is unreal' or that 'time does not exist' but because of memory and the placing of events in the past, present or future it cannot readily accommodate a world of *B*-succession which seems to confuse not just our view of time but everything to do with our experience of it. The modern, Russellian inspired, view holds that there is no such thing as a 'now' which moves. 'Time' in a *B* series alone, remains fixed and all relations in time are permanent. The belief is that though a static view of time makes us feel uneasy, logic demands that we take one. Schlesinger puts this down to: Russell's ingenious method of being able to talk about temporal experiences and yet eliminate reference to time,[66] the fact that science goes along with Russell's view, the vigorous defence given to Russell's view by other philosophers and that the 'transient view', as Schlesinger calls it, has not been argued for strongly enough. Schlesinger points out that whereas Zeno asserts that logic forces us to abandon spatial movement, Russell tells us that logic says there is simply no temporal position. Schlesinger believes that there are good reasons for those that hold to a transient view of time to 'launch a counter-offensive against the Russellians'.[67] Schlesinger uses Smart's argument against the transient view[68] as his example of the Russellian point of view and I follow this here.

Although Russellian *B*-theorists do not hold that we are immune from the psychological reasons for thinking that time moves, they claim that we are immune from the way that most people tend to think of time. Take, for example, Ivan Denisovitch going to prison in the Gulag for fifteen years. The *B*-theorist would say that there are two temporal alternatives to his incarceration. A week after his sentence began it would be possible to say his freedom is (un-consolingly) fifteen years away in one direction, which is the future, and also that it is (consolingly) only a week away in another direction, which is the past.[69] Ivan Denisovitch, therefore, has a level temporal choice about which view he takes and psychologically he would be more consoled by looking to the past. There is, however, no true consolation to be found in this. Common sense tells us that the past is no longer accessible because the temporal movement is only towards the future. The 'consoling' nature of the recency of a fixed past therefore means very little to us. Interpreting time like this is different from attaching an emotional state to our view about whether a bottle is half full or half empty. In spatial terms it is either half full or half empty. In asymmetrical temporal terms time is not either equally 'that way' or 'this way'. Whether we believe in a moving 'now' or not, a transitory view of time is the only sensible one in our everyday lives and there is nothing to be gained by imagining otherwise.

The *B*-theorist would respond, saying that the future problem of incarceration proving worrying was just the human psychology suffering under the illusion of time, and there is some truth in this. Indeed, McTaggart agrees that the speed of the passage of time varies (psychologically and actually) according to whether we are happy, sad, occupied or bored. If (ignoring the 'consoling' nature of the past) we accept this, then we could accept that Ivan Denisovitch's next fifteen years would go more quickly if he were well occupied. Accepting actual, experienced dilation of time because of the way we are involved in our experiences is a positive step away from the absolutist *Q* series.

If time goes at different rates because of our involvement in it, it would also be reasonable to believe that it is a quality distinct from space. Some *B*-theorists say that there may be an objective basis in treating time differently from space, time does, after all, seem to have a direction but space does not. Though, as Schlesinger points out, there is no reason space should not have a direction; it may be that we just do not know it yet.[70] However, unable to occupy himself, locate himself in any space other than his cell or to find any consolation in the fact that he has only been in prison for a week, the miserable Ivan Denisovitch may sink into wishing he was at a time before he was sent to prison; that is, to wish that his present was located somewhere at another point in the *A* series.

To wish oneself to be at a point in time ten years earlier would, on the *B*-theorist view, translate the statement of our present position from 'Now it is New Year 1995' to 'New Year 1995 is simultaneous with this utterance'. But psychologically, anyone wishing to be ten years younger would not be wishing that some utterance would be simultaneous with 1995. They would be wishing that instead of having consumed *n* years of their life they had consumed *n* minus 10 years. In other words they would want to be back in 1985. The *B*-theorist would point out that 'yearnings' and 'wishes' are incoherent (though no less frequent for their incoherence). It does not matter that 1985 was part of my experience (and therefore possibly available to 'return' to) because the incoherent desire to experience it in 1995 does not make it more available to experience. This sort of 'psychological matter' will only have a bearing on the nature of time if time is a psychological matter. My view is that time is in part a psychological matter. There is no empirical or psychological evidence that desires can transport us into the past. But this may be as much to do with the fact that the past does not exist as a thing in itself, as it is to do with the inadequacy of our desires to bring about temporal displacement. After all, the *B*-theorist allows that Ivan Denisovitch can be consoled by the fact that a week ago he was a free man and that this can have an effect in time, and if our degree of involvement in some experience can affect the apparent speed of the flow of time, it is only a step away from desires or wishes having some affect.

There are different ways of looking at the succession of events. Imagine that instead of going from 31 December 1994 to 1 January 1995 the succession goes from 31 December 1994 to 1 January 1985. In this universe the *B*-theorist would not admit that the now had jumped back to 1985, but that today, being the next day to 31 December 1994, happened to be an exact duplicate of 1 January 1985. Although everyone would think it was 1 January 1985, yesterday still would have been 31 December 1994. The calendar may say it was 1 January 1985 but the *B*-theorist still would say it was really the beginning of 1995 today. Schlesinger defends this view as equatable with space-time theory.[71] McTaggart's account of time (as an atemporal series) can fit in with this. Indeed, Schlesinger says, 'that Russell's account of what would happen if *P*'s wish came true is inferior to McTaggart's account'.[72] The *C* series is serial but oscillates. There is no reason that it cannot oscillate so that, in apparent time, it appears that a serial discrepancy like this has occurred. The *C* series has no fixed metric. Our perceptions (which are misperceptions of the *C* series) have a temporal order and the story that our perceptions tell us have an order in time. However, the order of the atemporal *C* series, though apparently a temporal order and the basis for our order of perceptions, can change. The *C* series is made up of perceptions and there is no regularity to this process.

Our common experience of time 'flowing' is inescapable. Common sense however, often has to give way to the truths of science. Smart says, 'Certainly we feel that time flows, but ... this feeling arises out of a metaphysical confusion'.[73] Schlesinger thinks this feeling is important. Time is part of our experience and he asks, 'Is it not an important rule that our account of the universe should be based as much as possible on experience?'.[74] For example, our assumption about the nature of solid objects is at variance with all physical theory. But perhaps this is not such a conflict as it at first appears. Science certainly can illuminate us about the nature of reality when our experience leads us to confusion, for example, in the case of matter. But so far no such thing has happened in the case of time; science has not shown us that time does not really 'move'.

If time does 'move' we need to know something about both the duration and the 'position' of the present. Again, Schlesinger gives us a useful perspective on this. He proposes a solar system, system *X*, in all respects the same as ours except that when we discover it, it is three hundred years 'earlier' than us and it's time is 'moving' at twice the rate of ours. That is, each day here is two days there. This means that someone in system *X* is not only 'younger' than us but is 'getting older' more quickly. This is based on the transient view that there are two times and not upon an experience of time like Ivan Denisovitch's. Schlesinger thinks there is no need for a hyper-time or hierarchy of other times to deal with this situation, he says, 'The existence of our own NOW is perfectly adequate for that purpose'.[75] Such a system not only provides for different 'nows' at different moments but for different speeds of 'nowness'. It is not how many events have occurred that dictates the objective speed of time here, it is the fact that system *X*'s time goes objectively faster than our own. A 'now' in system *X* is a faster 'now' than a 'now' in our system. Schlesinger proposes 'that the speed of time's flow is not a function of the number of events per unit [of] time'. Because the number of events in each of his proposed systems is more or less equal it is the speed of the 'now' which in *X* is twice as great.[76] We have here some essentials of Yourgrau's description of Gödel's rotating *R*-universe as ideal. We may call this the scientific investigation of temporal becoming (of which more in the next chapter). But back to system *X*. Schlesinger says, 'Since that situation [of the system *X*] is clearly not logically impossible it is false that time's flow or even the magnitude of the rate of its flow is in principle inaccessible to scientific study'.[77] That the situation is only imaginary does not in any way detract from its coherence as a description of time's flow. As MacBeath reminds us, 'The way in which the contingency of propositions about the topology of time is argued for is very often the argument from fantasy, which describes an allegedly possible world in which there are, for example, two temporally unrelated time systems'.[78] But this example does not translate into spatial terms.[79] It is sensible to ask what is an equivalent time in system *X* to 'now' in our system, but it is not sensible to ask what is an 'equivalent place' in system *X* to 'here' in our system. I use 'equivalent', Schlesinger just says, 'What time is it?' and 'What place is it?'. I think equivalent is more appropriate, though in either case I think there must be equivalence of both place and time between here and system *X* if we are to make any sense of the proposal.

If our own solar system is at the date 1995 and the other solar system is at a stage that more or less equates with 1695 in our solar system, then the events in the other solar system unfold at twice the speed of events in our own solar system. The speed at which, to us, things would appear to happen in the other solar system, would obviously be twice as fast as the speed of things that would happen on our own. On this view we cannot separate time and space. The idea that time-flow has no spatial counterpart is unsuccessful. Interestingly, Lowe reminds us that McTaggart pointed out[80] the possibility of the plurality of the *B* series.[81] Lowe also notes that Mellor points to a further difference in that events have spatial but not necessarily temporal parts.[82] Though Lowe feels that such a doctrine 'does not constitute a relevant difference between time and space'.[83]

Change is necessary in both system *X* and our own system. This view is therefore a tensed view of time in the *A* series. Schlesinger agrees with McTaggart that Russell's view of time cannot accommodate change. Schlesinger's interpretation of Russell's view is that of a time which stands still. MacBeath thinks this (Russellian view) should more appropriately be called 'the tenseless view of time'.[84] This division, between 'tensed' and 'tenseless' views, has led to a fresh analysis of McTaggart's partition of time into the *A* and *B* series (and fresh conclusions about why he is wrong). A brief synopsis demonstrates the circularity of these arguments and how attacking or re-analyzing McTaggart's terms does nothing to escape the conflict they create. Le Poidevin[85] agrees with Part 1 of McTaggart's argument, that is the reduction of the *A* series to the *B* series and that it is contradictory, but challenges Part 2, that the *B* series is insufficient to provide for temporal change. Based on this, he offers a tenseless theory of change. Following this, Lowe[86] attempts to show that McTaggart's paradox is fallacious.[87] Kiernan-Lewis,[88] referring to Prior,[89] says that the tenseless argument does not answer how we are continually compelled to believe that things really have ceased to exist. He thinks the tenseless view is not ontologically sufficient to deal with this. However, more recently, Kiernan-Lewis defends Prior's argument as both epistemologically and ontologically convincing.[90]

Maybe solipsism can provide us with an example of a view that logical analysis may consider acceptably 'realistic'. We would not need to abandon our assumption that we feel we know what happens in the minds of others. We would have to more fully appreciate the philosophical problems involved in trying to understand the meaning of other minds. *B*-theory accounts for our everyday experience, but offers no answer to the nature of time being at odds with our experience. The *B*-theorist thinks that the moving 'now' is not part of our direct experience; that it is an interpretation we must renounce in the face of the proof of logical analysis. McTaggart does not believe in the moving now either. He believes that the only possibility for time is in the *B* series. He finds the *B* series invalid because it relies on a self-contradictory *A* series and this brings him back to the same point as the *B*-theorist. Neither of them accepts the reality of time.

If we believe in the *A* series, the fact that all time seems to end in the past makes the past a strong possibility. But the apparent reliance that past has upon our memory of previously present events goes against the idea of an objective past. If time and space are unified, we must face a different set of possibilities. The 'now' and the 'past' may be linked, not just by time's flow but by their positions in the universe. If this is so the directional *A* series becomes a nonsense, for although the phases may resemble something in time they cannot be serial. However, the experience of the present is apparently in the present, and the present is the 'now' where everything in our experience actually happens. We cannot lay claim to actually 'experiencing' the past, whereas we continually believe that we have some access to the present.

The specious present is what we are presently experiencing and therefore consciously 'know'. The concept of 'present experience' is a tricky one. It seems that everything we can claim to know is actually in the past by the time we come to 'knowing' it (assuming that 'knowing' occurs in the brain and there is a time-delay between 'external' sensing and 'internal' knowing). But psychologically, present experience is essential, not only to our mental well-being but as a guarantor of our present witnessing of the objects of our experience.

We could say that some of our experiences are 'self-authenticating'. Pain for example would seem to be part of the self's present experience. But thinking this assumes the experience is part of the self and therefore immediately known and this is not so. When I say 'I have a pain', I do not mean 'my self has a pain'. My 'self' may indeed be 'in pain' because it is experiencing knowing about a painful experience but this is separate from the somatic experience of my body's pain. Because it is separate (or delayed) in this way it is later than the experience and the body's experience is in the past of the self's present experience of knowing. 'Knowing it as soon as I saw it' may seem closer to an immediate experience but this too is in the past. Such an experience is strictly a reflection; it is the self reflecting on the seeing and the knowing and combining them as near enough coincident. In fact, so nearly coincident that we can think our judgement is simultaneous with the experience. Indeed, reflecting that it *was* simultaneous gives us an exciting sensation of some sort of immediate knowledge and, as well as adding weight to the idea that they are simultaneous, also fits with what was most probably startling or shocking. It remains, however, that knowing followed the seeing and by the time we 'knew it as soon as we saw it' it was clearly in the past.

Common sense tells us that unless we know what is 'present' in our experience we can lay small claim to knowing anything. But most of what we are interested in takes longer than we can pin down to the specious present. We may, like St Augustine, as well as being despaired by time, wonder about how long time is. We may accept that there can be traces of past things or anticipation of future things in our minds at any present moment. From this we can suspect that these are indications of the reality of future and past. However, when it comes down to it — when it comes to deciding where it is that anything can happen — we conclude that the only place where anything can happen is in the present.[91]

But it remains that nothing that involves consciousness of an external stimulus is strictly instantaneous to us. The usual way around this is to 'extend' the present and make it 'thicker', so that whatever appears instantaneous also carries something of the past. This usually means talking of our present experience of time as having duration which, although a present experience, extends beyond the strictly instantaneous. To this end, Prior supposes a continuum of phases of change[92] in much the same way as Keeling, who, like Bergson, thinks the reality of time is to be found in its duration.[93] Our perception of the psychological disturbances involved in an 'instantaneous' experience dies down slowly. Our perception of an event appears to take up a greater period of time than the actual occurrence of the event. On Russell's example, our perception of a flash of lightning causes us to perceive something of a much greater length of time than the phenomena itself.[94] The *B* series is essential for this process and we can neither perceive a change nor remember anything without it. Indeed, although it is less important, 'immediate expectation' should also be considered part of perception. The past can only be understood in respect of our experience of the specious present and our specious present only has meaning if its duration extends into the past. We only understand that which succeeds, and anything that succeeds the specious present must be in the past. This would be so whether the specious present is objective (as Russell thinks[95]) or subjective. In this way, specious existence either cannot detect true temporal presentness or, if time is dependent (at least in part) upon mind, then presentness is undeniably entwined with, if not part of, the past.

Still, this does not resolve our feeling that we are experiencing a present moment, which itself, has a duration of presentness. There must be something happening that causes this experience. Maybe, as Broad says, it is to do with the way that we 'prehend' facts about the world. He rejects the specious present as an incorrect inference from the way we think. Indeed, he says that McTaggart wrongly always assumes that in the case of time there is some 'generally accepted doctrine on the subject'. Broad thinks such an assumption 'dangerous and unjustified' for he has never seen 'any account of the Specious Present which seemed even *prima facie* intelligible', considering it 'a verbal trick for evading ... difficulties'.[96]

According to Broad, presentness seems to be instantaneous but things that are instantaneous do not persist without qualitative change. If something present is perceived, it must also be instantaneous and therefore does not involve qualitative change; change cannot occur in an instant that is only occupied by an unchanging event. It is the case, however, that we do perceive objects that do persist as either changing or unchanged. So, although a perceiver is aware of only an instant of any durational period that at any one time is 'present' to him, there must be a perception of something beyond that instant. If this were not so there would be no apparent perception of change and this seems nonsense. According to Broad, the answer lies in us taking into account various amounts of past and interpreting them as specious presents. He says this psychological characteristic 'is capable of various degrees from zero up to a maximum' but incapable of prehending future.[97] On his 'wedge theory' he assumes that in approximate terms, any one mind's specious presents are of the same duration. That what is presented to that mind in all specious presents, as a maximum on the scale, will always be the same and that the degree of presentness reduces from maximum to zero at a uniform rate. He represents his scheme using a triangulation system in some respects not unlike a space-time cone. On this basis, any phase of time can be perceived as a temporal whole. The degree of presentness that is perceived will tail off from a maximum at its later boundary to a minimum at its earlier boundary. Because of this tailing off, the degree of presentedness will accordingly diminish. Perception of any temporal whole only ever takes place in the past. The perceived phase is completely past when perception of it first begins. The notion of temporal transition derives from our experience, in this way, of the specious present. The tailing off of presentedness from earlier to later, the continually overlapping specious presents and the reduction in perception of any one past, all conspire to form our apparent (though incorrect) experience of 'now'.[98]

Any specious present (and Broad's is a good version because it accentuates the pastness of apparent perceptions) is a description of 'temporal becoming' and, as such, imagines the flow of time very much as it appears to our intuition. Time is either flowing past us or we are passing though it (though the former is the usual belief). Either way, it is a fiction, because without us to experience it, there would be no such thing as a specious present. The notion of the specious present is a way of describing the part our consciousness plays in our experience of time (real or imagined). If we are completely sceptical about the objective world, then time may just be an expression of consciousness. Consciousness, in the Cartesian sense, may experience the presentness (or more correctly 'presentedness') of its own experience, but if it cannot experience the external world, then the external world (and time) is in doubt.

If there is no future (and the existence of future is particularly doubtful) then it must be the present that produces material for the past. Broad says, 'the essence of a present event is, not that it precedes future events, but that there is quite literally *nothing* to which it has the relation of precedence'.[99] McTaggart believes that we perceive only what we experience in our specious present.[100] If we agree, then whatever is earlier or later than this perception is not perceived at all but only judged to be in the past or future. The specious present, therefore, judgementally separates the future and past. If the time-series is to be objective then whatever is simultaneous with anything present must be itself present. If our perceptions of the time-series are supposed to be perceptions of an objective time-series then the specious present must accord with the objective present. But if there is another percipient then the length of the two specious presents may be different. This would mean that one percipient's specious present may overlap into the past of another. If one specious present lasts longer for one than for another, one still will be experiencing things in the present that for another, has become part of the past. For time to be real, the *A* series must exist but if this were the case, as McTaggart says, it could not exist objectively.[101]

There are some, however, who do not accept this. According to Swinburne, time may not have, 'of logical necessity a "unique topology"'.[102] MacBeath, however, thinks that Swinburne does hold to a 'standard topology' of time, considering it: unique, open, infinite, continuous and one-dimensional,[103] and proposes his own fantasy world in which time could be two-dimensional and retain the *A* series.[104] In the world of Zoë and Adrian

(MacBeath's reworking of the original story by Thomson), both were born on 1 January 1901. Zoë died on 1 January 1981 and Adrian died on 1 January 1941. From Zoë's point of view Adrian's life passed in fast motion. By the time he died at the age of 40, he was wizened and grey. Zoë went on for another 40 years until she died at the age of 80. From Adrian's point of view Zoë's life also passed at high speed. She died at 40 and Adrian went on for another 40 years until he died at the age of 80. There is no reason to think that the life of either of them was shorter than the other, only that their life-rates can be judged differently according to each's perspective of the other. It may be said that their lives were the same length if judged from some unknown measure or perspective. MacBeath illustrates this by two life-lines drawn at 60° from an axis that can be rotated inside a temporal framework. The life-lines can be the same length and yet cover different lengths of time in the framework.[105] Not surprisingly, this presents communication problems. Communication between two people with apparently the same perspective (and yet not the same perspective) leads to each not being in a position to attribute any utterance correctly.[106] Answers in a time-reversed world like this, will always be received before the question has been asked. Communication in the normal sense is not happening. Not only in the normally requisite temporal channel available (things are happening back-to-front), but because it is impossible to attribute the requisite condition of intention. Lucas argues that it is impossible for people to communicate in reversed time[107] but MacBeath disagrees.[108] He thinks that if Zoë can somehow come to know Adrian's perspective on her life the apparent problem of reversed-time communication is not so important. If (as Thomson and MacBeath suggest) Adrian kept a diary of events over a long period of time and Zoë could somehow come to read it, though she would be reading about her future, these future events would turn out exactly as Adrian's diary recorded them. She would, in this way, understand Adrian's perspective on her life and communication would have happened. This proposes a two-dimensional, serial temporality where simultaneous events are simultaneous according to their dimensional status.[109]

I think this temporal world is at least three-dimensional: the Adrian-dimension, the Zoë-dimension and a further dimension in which events could be judged simultaneous in both the Adrian-dimension and the Zoë-dimension. MacBeath initially envisages Adrian and Zoë in the same room but realises that this introduces an extra consideration; that, 'the room will have to have an interesting history of its own, quite apart from the history made by its occupants'.[110] His response is to separate Adrian and Zoë by first putting them on different planets, then in different galaxies. In order for

Adrian and Zoë to communicate in the 'full-blown' sense, they would have to invent a system for answering questions before they were asked. This is possible (according to MacBeath) by a process of sifting previous answers against facts that then come about and thus 'prove' them.[111] But this is a doubtful hypothesis; it is like forecasting the winner of a horse race. Still it is a forecast no matter how close the horses are to the winning post when the bet is placed. The forecast is based on anticipation and accords too much importance to specious existence. It extends the specious experience so that its 'edges' can overleap into past and future enough to pick up intentions held by someone with a different specious present. If this is looked at closely, we would have so many specious presents that the *A* series would collapse. MacBeath recognises that the communication problem cannot be accommodated by only two dimensions nor by physics and that it anticipates only a more complex world of more dimensions and that this would make the problems even worse.

But there is another, I think more confounding, problem with MacBeath's argument; his acceptance that 'communication' is synonymous with the maximum signal transmission speed — the speed of light. Apart from the dubious 'fact' that the speed of light is in any way synonymous with any maximum signal transmission speed (which I think is doubtful), he does not imagine, for example, that there is a possibility of 'simultaneous' communication. If self-awareness is a form of instant perception then there is no reason why awareness of another could also not happen instantly. On the other hand, if awareness of the external world suffers a perceptual 'delay', then why confine the delay to the speed of light?

According to McTaggart the passage of events through the present cannot be determined as simultaneous with a specious present. Any duration accorded to the present must be fixed independently of any perception. If not, we are faced, as we have seen, with the problem of the *A* series in one percipient's present containing another percipient's past; and this is not borne out by experience. We can throw out the idea of a present 'point' for this is not inkeeping with our experience of events that have variable duration. McTaggart says, 'The time in which we experience them [events] has a present of varying finite duration, and is therefore divided into three durations — the past, the present, and the future'. It is, therefore, not proper to confuse the term 'present experience' with the idea that there is an objective present point. McTaggart thinks that the idea of all objective time that only allows the present to be a point where the difference between future and past is revealed, 'has nothing but the name in common with the present of experience'.[112]

The present is our most 'real' experience of time and yet it cannot stand much analysis. If the present is in doubt, we must begin to wonder, like McTaggart, that 'the denial of the reality of time turns out not to be so very paradoxical'.[113] I think time, as we experience it in the specious present, is an illusion. Still it would be an illusion even if our experienced 'realities' actually existed. All our experience of time is in the specious present but this experience can never correspond with an external 'present', whether it is real or not. As McTaggart says, 'For if time is unreal, nothing can be in any present at all, and, if time is real, the present in which things are, will not be a specious present'.[114]

I think, like Keeling, that the true nature of what we misperceive as specious temporality lies in the nature of change. He believes that a period of time has magnitude that is successive. This magnitude is constituted of duration that is an ingredient in any change in an enduring thing. Duration is in itself a quantitative and unified whole. Change is known by the different phases undergone by a changing thing and these phases 'determine' temporal periods. To understand change we need to differentiate between duration — a lapse of time accorded to an existent which proceeds, and the duration of the existence of a thing as a whole. In other words, between the duration of part of the 'life' of a thing and the duration of a thing that is its whole 'life'. The latter category is a thing that is no longer proceeding or is in the past. Because the diverse temporal periods could be determined by the changing of the continuant existent throughout the whole of that continuant's existence, this leads to the distinctions past, present and future being found in the characteristics of phases of change. Although the existent endures in continuance, such endurance is not temporal, though it is temporally determinable. There is, therefore, no series but a whole course of change.

Such a view allows us to reject an absolutist view of time and change and put in its place a pluralism of courses of change. Because courses of change have a certain quantity, as Spinoza supposed, they can be measured and this leads us to our mistaken view of time. Based on this, we can deny the logical priority of time to change, the duality of time both 'passing' and being that 'which passes', serial time and the idea that time or change could be substantival existents.

We then have a world in which temporality is a logically prior, non-temporal endurance of existents that reveals quantitative phases of change as they become repeated. A 'temporal period' is a function of a constant (the continuant's existence) and two variables, the characterisation of its state and of its phase of changing. This is a world where the present, as the domain of change, is the only reality and the only place where action can occur. The successive renewal of presentness is the ultimate significance of what we mistakenly believe in as time. Thus, by resolving the problem of presentness, we are again faced with having to accept that time is unreal. Present, like past, has failed any attempts at sufficient description. Now, we pass on to the third phase of the *A* series.

4.3.3 Future

The future is the source of the river of time and its most unacceptable element. As we have seen it is possible to retain the *A* series while denying that future exists.[115] If we take this view, any judgements we make about the future do not refer to any facts at the time that they are made. Future facts cannot therefore be true or false. When there is a fact for us to refer to, it must be in the past and there remains either true or false. As Broad rightly says, judgements about the future only 'enjoy a courtesy title by anticipation, like the elder sons of the higher nobility during the lifetime of their fathers'.[116] The future is, therefore, not really an existent. That the future is non-existent also has important psychological consequences. A non-existent future is supported by those who argue that only if the future (as that which will come later) is 'empty' can there be any meaning to freewill (not least of these is Bergson). Bergson imagines the future to be undetermined at the present moment; it is a reality in which 'something unforeseeable and new' is continually being created.[117]

Let us, however, imagine that the future exists. If it does, then either it produces events for the present or the present somehow produces the events of the future. If our actions in the present influence the future then we alter the serial nature of the *A* series which we assume to be transitive and asymmetrical. According to Aristotle, we cannot breach this asymmetry; there is or there is not going to be a sea battle tomorrow.[118] No matter what we do to try to influence this future event we are acting in vain. The future is already made, even if 'made' of alternatives. However, the inescapability of a future fact that is either one fact or another fact, raises a problem for most of us. We are not comfortable with the idea that the 'certainty' from which there is no escape is made up of one fact *or* another fact. Anscombe thinks this discomfort is because 'Aristotle's considerations aren't about knowledge and certainty, and *are* about necessity and its corresponding possibility'.[119] To put it another way, we may think that one or another fact will occur but still feel that the future is linked to the succession of events that we have so far experienced.

Uncertainty like this leads us to believe in a contingent future. Events that occur in our recurring present change, and do not repeat events that have occurred previously. They must, therefore, 'lie' somewhere else and this place we call the future. This shifting of presentness into the immediately (almost) future presentness is unconvincing and is what Spinoza calls a 'wavering of the imagination'.[120] But we could never be certain that future present was made up of new present events. There is no reason why previously past events should not start occurring again without us knowing it. However, if they did it would only be for the mark II version or some strange memory-blocking repeat. Although their happening might be contingent, the contents of the events would not be because they would be contingent on whatever they were when they occurred previously.

The future may be the 'possibility' that becomes known in the present; what Lucas calls the 'passage from possibility through actuality to necessity'.[121] This view of future is again like Aristotle's, it is 'open' but 'unalterable'. It somehow 'contains' or 'holds' all the possibilities that we may, in the present, imagine or not imagine. They will only be known when they occur in the present. However, when they do occur, they could only have occurred by having previously been available as a possibility, and this is proved by the fact that now it has occurred, as opposed to would or would not occur.

I think this close interconnection between present and continually occurring 'used' future can lead to further error. It can cause us to believe that, although the past is something unalterable, our actions in each subsequent present can influence and change what will happen in the future. This presupposes that the future already exists, waiting to be either discovered or used up. It assumes that it is there, not just as an emptiness to be filled by present action, but in some way already having integral potential occurrences (or at least pre-dispositions). This is an incorrect view which is caused simply by wrongly concluding that because we experience a continual flow of present that the present is fed from the future. Because we do not know where this stream of presents comes from should not cause us to believe they come from something like the future.

Nevertheless, such a future has many uses. We can use it pragmatically, as James thinks, to test our current beliefs[122] or even use it to foretell what is going to happen. After all, there is little point in making a weather forecast unless we believe there is a future from which it will occur. Dunne believes that we can gain access to this sort of future while still being in the present.[123] But if the future is sitting there waiting now, then it must also be present. We may not be experiencing it yet, but relative and synchronous to our now, it exists.

Relativity theory subsumes all possible future events into an amalgam of future, present and past events that are known, or not known, according to our place and velocity in the universe. In such a universe there cannot be a real modal or ontological future or past. Such futurity is dependent upon our present reference point as one observer's future may be another observer's past. A determined event may only be a determined future event for one observer. For all others, the same event could be in the past or present. This does away with the confusing 'wavering' we experience when we think of future somehow converting itself into present or the imponderability of whether the future is or is not, or contains anything or does not. All events could be present or past, and future is then simply a way of describing events we do not know. But the fact that we do not know events and that continual present causes us to know more events, throws us back onto the experience of something that involves future.

The relevance that future events have to us as individuals is affected by how far in the future they are placed. For example, 'It is my birthday in one week's time' is relevant; 'It is my birthday in nine months time' is relatively relevant; 'It is my birthday in nine years time' is relatively irrelevant and 'It is my birthday in nine hundred years time' is irrelevant. Repeated events like birthdays, have different 'places' in the future. Their relevance passes from the most relevant, 'It's my birthday tomorrow', via the rather absurd though probable, 'It's my birthday in nine years time', to the irrelevant, 'It's my birthday in nine hundred years time'. It is most improbable that I will be alive in nine hundred years time. Whether or not, in nine hundred years time, 'my birthday' still would be 'mine' is questionable. Whereas, barring ill-fate, I could lay claim to my birthday 'tomorrow', I could not lay claim to my birthday nine hundred years from now. This sort of future only contains potential events that only have relevance to our own lives or social conduct.

Our 'knowledge' of future events is dependent upon how those future events work out. If the prediction that we make comes true then we could say, with some sincerity, that we knew all along that it would work out. If, on the other hand, it does not work out then we would have to concede that we were wrong to assume that we knew in the first place. This 'retrospective withdrawal proviso',[124] makes the future precarious to a greater or lesser degree. In the first place it either does or does not hold a certain fact. In the second place, if we predict that fact (at the time of the prediction not knowing whether it will match our prediction or not) and when it actually occurs, our prediction proves to be correct, then, at this point only, can we say that our prediction was in fact a 'former knowledge'. This former knowledge is only available to be called a 'former knowledge' when we know the outcome. This is an outcome of the 'prediction', as it was called, before we knew the outcome. When we made the prediction, the future was not known. When the event in the future actually occurs or does not occur, that previously future event becomes known in respect of the previously 'prior to that future event' point of view. This view of future uses the present view of the past — taking a point of view from the past — to derive a particular point of view of the future, which is now the present.

But there is no epistemic certainty in the future unless I can truly predict. For example, I may say that I 'know' that I am going to die. Such a statement seems certain enough, indeed it seems to be the most certain prediction about the future. Yet, if I were the first immortal human or suddenly, from this moment, the universe embodied a condition in which no living thing died, my prediction would have been hopelessly wrong. When I arrived in the future to find out how wrong I had been, there would be no sense in saying that the prediction was made 'according to the prevailing conditions at the time', for those conditions no longer prevail. Only if, we truly know what is in the future, and this is consequently borne out, can we 'know' the content of the future. This way of 'truly knowing the future' is not available to us. To say that there is an acknowledged succession of events that have so far been correct and are therefore sufficient for us to know how the future will turn out, may be good enough for ordering our lives but it is not good enough for knowing the future.

For McTaggart the idea that the future exists because it contains the possibility 'is' or 'is not' is irrelevant to the argument that the future exists. If the universe ended in five minutes time, it would be an ill-fated argument that said we had decided that in ten minutes something would or would not exist. The possibility that there may not be any more future five minutes from now only further weakens the argument for the *A* series. Although the *A* series can exist without future we would be hard pressed to sort out what 'fed' the present in its absence. Even so, the fact that future cannot be known until it becomes present should cause us to suspect that there is indeed such a thing as future, and that we have produced the term only as a means of explaining away the puzzling re-occurrence of 'more' present.

4.4 Conclusion

I conclude that none of our commonly held views give us good reason to accept the *A* series. In § 4.2 we saw that although we can take a variety of views about time, none of them is positively conclusive. We saw that the idea of the non-existence of time encourages us to think that it cannot be 'time' that does not exist and whatever is happening that appears as time must be something else. Even so, we found that 'common sense' is something worth holding onto; it is a useful measure against which we can lay the riddles of our experience of temporality. Having looked more closely at the *A* series in § 4.3 (which, based on McTaggart's view is essential for the reality of time) the matter only seemed to get worse. In § 4.3.1 we found that although the concept of past has important social and psychological consequences, as a separate phase of change it shows itself inaccessible to coherent description or proof. In § 4.3.2 we saw that present, like past, fails any attempt at sufficient description. In § 4.3.3 I concluded that the fact that future cannot be known until it becomes present causes us to suspect that there is indeed any such thing as future, and that we have produced the term only as a means of explaining away the puzzling re-occurrence of 'more' present. Although the *A* series remains an inescapable part of how time appears to us, analysis of its phases produced various conflicting and unconvincing results. Although it could be that this sort of analysis of the *A* series is misleading; it is possible that time will not yield to this sort of scrutiny and this, by itself, does not mean that it is unreal. But this is improbable. If time in the *A* series is real then its independent phases should be credible. We have seen that this is not so and should not expect to find any greater sense from them if we bundle them together. The *A* series certainly provides us with a convenient everyday mechanism for ordering our lives but its inherent inconsistency does nothing to vouch for its (and consequently time's) reality. The *A* series is not itself an escape route from its own inherent problems nor a consistent system for temporal ordering.

I conclude that it is unsatisfactory that no conventional view provides an answer to some natural questions we have about time's nature. I take the view that on rational and common sense grounds we are justified (like McTaggart) to look outside its terms for an answer.

Notes

1 Le Poidevin and MacBeath, *The Philosophy of Time*, p. 1.
2 See P. Radin, *Primitive Man as Philosopher* (New York, Dover, 1957).
3 For a full discussion of this see Whitrow, *Time in History*, pp. 21-9.
4 See Shallis, 'Time and Cosmology', p. 63.
5 For a straightforward exposition see A. Einstein, *Relativity: the Special and the General Theory: A Popular Exposition*, trans. by R.W. Lawson (London, Methuen, 1988. First published 1920).
6 St Augustine, *Confessions*, ed. W.H.D. Rouse, trans. by W. Watts (London, Heinemann, 2 vols., 1912), vol. 2, Bk. 9, ch. 22, p. 237.
7 D. Hume. *A Treatise of Human Nature*, ed. and with an Analytical Index by L.A. Selby-Bigge, 2nd edn., revised by P.H. Nidditch (Oxford University Press, 1978), p. 33, § 3.
8 G. Ryle, *The Concept of Mind* (Harmondsworth, Peregrine, 1983. First published 1949) p. 154.
9 For extended discussion on these two strands, see J. Earman, *World Enough and Space-Time* (Cambridge, Mass., MIT Press, 1989).
10 F. Cajori, *Newton's Principia*, trans. by A. Motte (Berkeley, Calif., California University Press, 1934), p. 6.
11 Russell, *The Principles of Mathematics*.
12 G. Berkeley, *De Motu*, in *Philosophical Works including the Works on Vision*, eds. A.A. Luce and T.E. Jessop with Introduction and Notes by M.R. Ayers, trans. by A.A. Luce (London, J.M. Dent, 1985). *Of Motion* is a full discussion on this, pp. 211-27.
13 E. Mach, *Science of Mechanics*, trans. by T.J. McCormack, 6th edn. (Chicago, Open Court, 1960), p. 284.
14 D. Sciama, 'Time "Paradoxes" in Relativity', in Flood and Lockwood, *The Nature of Time*, p. 14.
15 M.F. Cleugh, notes on her personal copy of J.W. Dunne, *An Experiment with Time* (London, Faber and Faber, 3rd edn., 1934), p. 141.
16 See Gunn, *The Problem of Time, An Historical and Critical Study*, p. 323.
17 B. Russell, *Human Knowledge. Its Scope and Limits* (London, George Allen and Unwin, 1948), p. 287.
18 McTaggart, *The Nature of Existence*, § 306.
19 I. Kant, *Critique of Pure Reason*, (1781; 2nd edn., 1787), trans. by N. Kemp Smith (London, Macmillan, 1973), p. 75.
20 Lucretius, *De rerum natura*, in *Lucretius: on the Nature of the Universe*, Introduction and trans. by R.E. Latham (Harmondsworth, Penguin, 1951), Bk 1, lines 455-65, pp. 40-1.

21 See G.W. Leibniz, *Philosophical Writings*, ed. and Introduction by G.H.R. Parkinson, trans. by M. Morris and G.H.R. Parkinson (London, J.M. Dent, 1988), pp. vii-xix.

22 Leibniz, *The Monadology*, §§ 82-90.

23 Leibniz, 'Third Paper to Clarke', in *Leibniz, Philosophical Writings*, p. 212.

24 McTaggart, *The Nature of Existence*. Summarily in § 903.

25 H. Reichenbach, *The Philosophy of Space and Time*, trans. by M. Reichenbach and J. Freund (New York, Dover, 1958), p. 287.

26 Gödel, 'A Remark about the Relationship between Relativity Theory and Idealistic Philosophy' in Yourgrau, *Demonstratives*, pp. 261-5.

27 See J. Dalton, *A New System of Chemical Philosophy* (Manchester University Press, 2nd edn., 1808).

28 For early discussion on types of particle see E.S. Eddington, *The Nature of the Physical World*, Gifford Lectures, 1927 (Cambridge University Press, 1928), p. 201. For discussion on Quantum mechanics see P. Davies, 'Time Asymmetry and Quantum Mechanics', in Flood and Lockwood, *The Nature of Time*, p. 122. For a more recent view of the building-blocks image of the universe see J. Powers, *Philosophy and the New Physics* (London, Methuen, 1982), p. 160.

29 Keeling, *Time and Duration*, p. 27.

30 See Einstein, *Relativity. The Special and the General Theory.*

31 W.H. Newton-Smith, *The Structure of Time* (London, Routledge and Kegan Paul, 1980), p. 195.

32 This was implied by Maxwell (J.C. Maxwell, *A Treatise on Electricity and Magnetism*, (Oxford, Clarendon Press, 2 vols., 1873)) and demonstrated empirically by Michelson (A.A. Michelson, 'On the Relative Motion of the Earth and the Luminiferous Ether', *The American Journal of Science and Arts*, vol. 22 (1881), no. 128, pp. 120-9). Cedarholm and Townes confirmed that there was no fixed ether through which the solar system was passing. (J.P. Cedarholm and C.H. Townes, 'A New Experimental Test of Special Relativity', *Nature* (1959), no. 184, pp. 1350-1).

33 See M. MacBeath, 'Times Square' in Le Poidevin and MacBeath, *The Philosophy of Time*, pp. 183-202.

34 See Gödel, 'A Remark about the Relationship between Relativity Theory and Idealistic Philosophy', also P. Yourgrau, *The Disappearance of Time: Kurt Gödel and the Idealistic Tradition in Philosophy* (Cambridge University Press, 1991).

35 H.H. Price, 'The Appeal to Common Sense', *Journal of Philosophical Studies*, vol. 5 (1930), no. 17, p. 24.

36 See Bergson, *Creative Evolution.*

37 Price, 'The Appeal to Common Sense', p. 33.

38 H. Reichenbach, *The Direction of Time*, Berkeley, Calif., University of California Press, 1956).

39 L. Sklar, 'Up and Down, Left and Right, Past and Future', in Le Poidevin and MacBeath, *The Philosophy of Time*, p. 109. First published in *Noûs*, vol. 15 (1981), no. 2, pp. 111-29. See also Sklar's essay, 'What Might Be Right about the Causal Theory of Time', *Synthèse*, vol. 35 (1977), no. 2, pp. 155-71.

40 J. Earman, 'An Attempt to Add a Little Direction to "The Problem of the Direction of Time"', *Philosophy of Science*, vol. 41 (1974), pp. 15-47.

41 M. Dummett, 'Causal Loops', in Flood and Lockwood, *The Nature of Time*, pp. 135-69.

42 P.J. Riggs, 'A Critique of Mellor's Argument against "Backwards" Causation', *British Journal of Philosophy*, vol. 42 (1991), pp. 75-86.

43 See Mellor: *Real Time*, pp. 160-87; 'Fixed Past, Unfixed Future', in B. Taylor, ed. *Michael Dummett: Contributions to Philosophy* (Dordrecht, Nijhoff, 1987), pp. 168-88; 'Causation and the Direction of Time', *Erkenntnis*, vol. 35 (1991), nos. 1/3, pp. 191-203.

44 A. Quinton, 'Spaces and times', *Philosophy*, vol. 37 (1962), no. 140, pp. 130-47). A fuller exposition of his idea is found in A. Quinton, *The Nature of Things* (London, Routledge and Kegan Paul, 1973).

45 R. Swinburne, 'Times', *Analysis*, vol. 25 (1965), no. 108 (NS), pp. 185-91.

46 R. Swinburne, *Space and Time* (London, Macmillan, 1981), pp. 205-6.

47 M. Hollis, 'Times and Spaces', *Mind*, vol. 76 (1967), no. 304, pp. 524-36.

48 Swinburne, *Space and Time*, ch. 3.

49 Broad, *Scientific Thought*, p. 66.

50 McTaggart, *The Nature of Existence*, § 341.

51 K. Gödel, 'A Remark about the Relationship between Relativity Theory and Idealistic Philosophy'.

52 One of the most popular books in this field, an essentially mathematical discussion with metaphysical conclusions, is S. Hawking, *A Brief History of Time* (London, Bantam Press, 1988). For an approach which integrates science, computing and philosophy see D. Deutsch, *The Fabric of Reality* (Harmondsworth, Penguin Press, 1997). A standard text on general relativity is, C.S. Misner, K.S. Thorne and A. Wheeler, *Gravitation* (San Francisco, W.H. Freeman, 1973).

53 McTaggart, *The Nature of Existence*, footnote to § 311.

54 Dummett, 'Causal Loops'.

55 Aristotle, *De Anima (On the Soul)*, trans. by J.A. Smith in *The Basic Works of Aristotle*, 415b 12.

56 Yourgrau, *The Disappearance of Time*, p. 147.

57 St. Augustine, *City of God*, trans. by J. Healey, Introduction by E. Barker (London, J.M. Dent, 1931) Bk. 10, Ch. 26.

58 See also: J. Harrison, 'Dr Who and the Philosophers', *Proceedings of the Aristotelian Society*, vol. 45 (1971), pp. 1-24; M. MacBeath, 'Who was Dr Who's Father?', *Synthèse*, vol. 51 (1982), no. 3, pp. 397-430; P. Horwich, 'On Some Alleged Paradoxes of Time Travel', *Journal of Philosophy*, vol. 72 (1975), no. 14, pp. 432-44; W.L. Craig, 'Tachyons, Time Travel, and Divine Omniscience', *Journal of Philosophy*, vol. 85 (1988), no. 3, pp. 135-50; C. Ray, *Time, Space and Philosophy* (London, Routledge and Kegan Paul, 1991), ch. 8.

59 Loisou, *The Reality of Time* (Aldershot, Gower, 1986), pp. 157-67.

60 Ibid., pp. 158-60.

61 Findlay, 'Time: A Treatment of Some Puzzles'.

62 Prior, *Past, Present and Future*, p. 1.

63 McTaggart, *The Nature of Existence*, § 311.

64 See Thucydides, *The History of the Peloponnesian War*, ed. and trans. by R. Livingstone (Oxford University Press, 1960).

65 See Homer, *The Iliad*, trans. by E.V. Rieu (Harmondsworth, Penguin, 1985).

66 B. Russell, 'On the Experience of Time', *The Monist*, vol. 25 (1915), no. 2, pp. 212-33.

67 G.N. Schlesinger, 'How Time Flies', *Mind*, vol. 91 (1982), no. 364, p. 510.

68 Smart, 'The River of Time'.

69 Ibid., p. 511.

70 Ibid., pp. 511-12.

71 Ibid., pp. 522-3.

72 Ibid., p. 514.

73 J.J.C. Smart, 'Time and Becoming', in *Time and Cause: Essays Presented to Richard Taylor*, ed. P. van Inwagen (Dordrecht, Reidel, 1980), p. 3.

74 Schlesinger, 'How Time Flies', p. 515.

75 Ibid., p. 519.

See also G. Plumer, 'Why Time is Extensive', *Mind*, vol. 93 (1984), no. 370, pp. 265-70.

76 Schlesinger, 'How Time Flies', p. 519.

77 Ibid.

78 MacBeath, 'Time's Square', p. 184.

79 Schlesinger, 'How Time Flies', p. 521.

80 McTaggart, *The Nature of Existence*, § 323.

81 E.J. Lowe, 'The Indexical Fallacy in McTaggart's Proof of the Unreality of Time', *Mind*, vol. 96 (1987), no. 381, p. 62.

82 Mellor, *Real Time*, p. 111.

83 Lowe, 'The Indexical Fallacy', p. 63.
See also R. Le Poidevin and D.H. Mellor, 'Time, Change and the "Indexical Fallacy"', *Mind*, vol. 96 (1987), no. 384, pp. 534-8 and E.J. Lowe, 'Reply to Le Poidevin and Mellor', *Mind*, vol. 96 (1987), no. 384, pp. 539-42.

84 M. MacBeath, 'Clipping Time's Wings', *Mind*, vol. 95 (1986), no. 378, pp. 233-7.

85 R. Le Poidevin, *Change, Cause and Contradiction: A Defence of the Tenseless Theory of Time* (London, Macmillan, 1991).

86 E.J. Lowe, 'McTaggart's Paradox Revisited', *Mind*, vol. 101 (1992), no. 402, pp. 323-6.

87 See also Le Poidevin's response (R. Le Poidevin, 'Lowe on McTaggart', *Mind*, vol. 102 (1993), no. 405, pp. 163-70) and Lowe's reply (E.J. Lowe, 'Comment on Le Poidevin', *Mind*, vol. 102 (1993), no. 405, pp. 171-3).

88 D. Kiernan-Lewis, 'Not Over Yet: Prior's "Thank Goodness" Argument', *Philosophy*, vol. 66 (1991), no. 256, pp. 241-3.

89 A. Prior, 'Thank Goodness That's Over', *Philosophy*, vol. 34 (1959), no. 128, pp. 12-17.

90 J.D. Kiernan-Lewis, 'The Rediscovery of Tense: A Reply to Oaklander', *Philosophy*, vol. 69 (1994), no. 268, pp. 231-3.

91 St Augustine, *Confessions*, Ch. 11, paras. 23-4.

92 Prior, 'Changes in Events and Changes in Things', p. 38.

93 Keeling, *Time and Duration.*

94 Russell, *Human Knowledge*, p. 226.

95 Ibid., p. 232.

96 Broad, *Examination*, vol. 2, pp. 281-2.

97 Ibid., p. 282.

98 Ibid., p. 285.

99 Broad, *Scientific Thought*, p. 66.

100 McTaggart, *The Nature of Existence*, § 344.

101 In § 344 of *The Nature of Existence* the statement reads, 'And, since the *A* series is essential to time, this involves that the *A* series is subjective'. This should read 'objective' and was quite correctly picked up by Broad and noted in his *Examination*, p. lvi.

102 Swinburne, *Space and Time* p. 209.

103 MacBeath 'Time's Square', p. 184.

104 See most recently, MacBeath, 'Time's Square'. Also: J.J. Thomson, 'Time, Space and Objects', *Mind*, vol. 74 (1965), no. 293, pp. 1-27; J.W. Meiland, 'A Two-Dimensional Passage Model of Time for Time Travel', *Philosophical Studies*, vol. 26 (1974), nos. 3/4, pp. 153-73; T.E. Wilkerson, 'Time and Time Again', *Philosophy*, vol. 48 (1973), no. 184, pp. 173-7 and T.E. Wilkerson, 'More Time and Time Again', *Philosophy*, vol. 54 (1979), no. 207, pp. 110-12. This view is argued against forcibly in R.E. Nusenoff, 'Two-Dimensional Time', *Philosophical Studies*, vol. 29 (1976), no. 5, pp. 337-41 and R.E. Nusenoff, 'Spacialized Time Again', *Philosophy*, vol. 52 (1977), no. 199, pp. 100-1.

105 MacBeath, 'Time's Square', pp. 188-91.

106 Ibid., p. 191.

107 J.R. Lucas, *A Treatise on Time and Space* (London, Methuen, 1973), pp. 45-6.

108 M. MacBeath, 'Communication in Reversed Time', *Synthèse*, vol. 56 (1983), no. 1, pp. 27-46.

109 Ibid., p. 196.

110 Ibid., p. 200.

111 Ibid., p. 199.

112 McTaggart, *The Nature of Existence*, § 345.

113 Ibid., § 346.

114 Ibid.

115 Broad, *Scientific Thought*, p. 66.

116 Ibid., p. 73.

117 Bergson, *Creative Evolution*, p. 359.

118 Aristotle, *De Interpretatione (On Interpretation)*, trans. by E.M. Edghill, in *The Basic Works of Aristotle*, Ch. 9, 19[a], lines 30-34.

119 G.E.M. Anscombe, 'Aristotle and the Sea Battle, De Interpretatione, Ch. IX', in J.J.C. Smart, ed. *Problems of Space and Time* (New York, Macmillan, 1964), pp. 43-57, p. 56. A revised version, with some omissions and additions, originally in *Mind*, Vol. 65 (1956), no. 257, pp. 1-15).

120 B. Spinoza, *Ethics*, revised by G.H.R. Parkinson and trans. by A. Boyle (London, J.M. Dent, 1985), note to Prop. 44, pt. 2, p. 73.

121 J.R. Lucas, 'The Open Future' in Flood and Lockwood, *The Nature of Time*, p. 126. See also J.R. Lucas, *The Future* (Oxford, Basil Blackwell, 1989), in which he defends an open future as opposed to a future in a block universe.

122 See W. James, *The Principles of Psychology* (Chicago, Encyclopaedia Britannica, 1952. First published 1891). In particular Ch. 15, 'The Perception of Time', pp. 396-420 and Ch. 16, 'Memory', pp. 421-51.

123 Dunne, *An Experiment with Time.*

124 Lucas, 'The Open Future', p. 129.

5 Some Philosophical Positions on Time

5.1 Introduction

Philosophers, however, do not always accept everyday reasoning nor the voice of intuition. It now becomes necessary to mark out the philosophical ground. Doing this will show how McTaggart is placed in philosophical debate and will demonstrate that, philosophically, we have no alternative but to proceed to an analysis of atemporality. It will also show that McTaggart is not so far from either common sense or scientific belief as at first he seems.

I now look at some general philosophical positions on time that can be adopted in relation to our experience of it. Because we have found in Chapter 4 that precise answers to time's puzzling nature cannot be found implicit in the way we describe it, there is a chance that our experience of it, and the way philosophy and science have interpreted it, may provide something intuitively more positive.

Time does not give us access to its nature. This leads us to take certain positions that combine our views in ways that are satisfactory to us. These can be divided into three central philosophical positions. In § 5.2 I look at 'reality of consciousness plus being'. This is where we commonly expect to find an idealist like McTaggart. Here I show that there are good reasons to believe that McTaggart's fuller philosophy throws light on questions about time and has direct relation to the modern scientific view. In § 5.3 I look at 'reality of becoming plus consciousness'. The discussion here reveals how on this view there needs to be 'something' in which 'something happens' and this throws us back into the confusing nature of the *A* series. This negative conclusion, however, inadvertently gives McTaggart some support. § 5.4, 'a four-dimensional universe with no becoming', surprisingly puts McTaggart into a scientifically valid position that, at first sight, we might think hostile to his idealist argument. Here I show that idealist thinking, far from being superseded by modern scientific views, combines meaningfully with them and that McTaggart's atemporal reality can make a direct contribution to four-dimensional thinking. In § 5.5 I conclude that McTaggart's position is surprisingly close to the views held by science. It also appeals to some of our common sense values. For these reasons alone it is important to find out what he has to say on the subject.

5.2 Reality of consciousness plus being

McTaggart's argument has two separate theses. The first, the positive thesis, is his analysis of time that defines the *A* and *B* series. The second, the negative thesis, leads to his denial of the reality of time based upon the contradictory nature of the terms of the *A* series (that he considers essential for time to be real).

One group of analytic philosophers agrees with his first thesis and disagrees with his second. Another group disagrees with the first part and argues that time still can be real, even if his second thesis is correct (though the second thesis relies upon the first). Some (though these are few) hold that it is to McTaggart's greater scheme that we must look. It is in this context that we find the reason for apparent time being erroneous. It is here where the reality of consciousness, plus a special sense of being, debunks time altogether. The special sense of being is one used to denote a subject of consciousness (a self) and is therefore an entity to be contrasted with mere 'objects'. If time were real such a being could presumably experience temporal becoming.

We usually imagine the metaphor of the 'flow' of time in one of two ways. We can think of ourselves or events as moving from the past into the present and then into the future. Alternately we can think of ourselves as somehow fixed to the present and events coming from the future into the present and then into the past. Temporal becoming proposes a third alternative. Gale first uses the term to describe the idea of events progressively coming into being.[1] Again, there are two possible views of this. That it is the perceiver's view of events or that it is events themselves that come into being. Which of these we choose is important. Perception is necessary in both cases. If it is the perceiver's view of events, then the perception may be erroneous (which is McTaggart's view). If it is events themselves that come into being, then we must accept a material world, our perception of which may or may not be in error. The notion of the reality of consciousness plus being discounts this latter view.

The idea of consciousness plus being has its roots in the holistic tendency of continental philosophy and draws on Bergson's *la durée*, the time of consciousness. When Merleau-Ponty says that time 'arises from *my* relation to things',[2] he implies a movement towards the future of a conscious being. On this view there could not have been any time before it was consciously perceived. For although there was unperceived time prior to consciousness, it is only known because it is consciously perceived now. As such, time becomes a function of consciousness. Commonly understood time arises from conscious beings agreeing on understood temporal values. Any idealist notion we have of temporal becoming therefore relies on consciousness being part of a person's living presence. This anthropocentric view can be applied to both the *A* and the *B* series. Temporal becoming can be in the present of the *A* series or at the centre of the succession from earlier to later in the *B* series.

It is common these days to adopt the *B* series view. Mellor is a central exponent of this view, though of course the view existed much earlier, Russell being the pre-eminent *B*-theorist. The *B* series view supposes that it is things that change in an objective world. McTaggart's view is that it is not things that change but events that happen to things that change. He thinks that this change occurs in a spiritual world that has no material, objective status.

That the universe always seems to be in time is reliant upon perception or the changing nature of existence. Though this changing may only be apparent and therefore be a perceptual issue. Heller upholds the distinction between temporal direction and spatial direction. He thinks that we can have a perspective of spatial direction but not of temporal direction. The asymmetry of time dictates that things have different properties at different times and therefore must change.[3] Lombard argues against this.[4]

It is not possible to account for any true temporality based simply upon changes in the existent. If unobserved changes occur, which is perfectly reasonable to suppose, then those changes cannot be part of an *A* series as we know it. The future could well be the past and the past the future, the 'direction' of time would be unknown. After all, our 'beginning, middle and end' attitude about our lives, and therefore the directionality of time, may conflict with the apparent movement towards increased entropy. If we accept that entropic direction is a universal indicator then time could be 'moving' in the opposite direction to how we commonly view it. Without observed change there would never be any present because present relies upon the relational stance of the observer. If unobserved change could occur, it could not occur in the *A* series but nor could it occur in the *B* series. The *B* series relies upon the *A* series. Without an observer a *B* series alone would suffer all the same problems as the *A* series. Earlier could be later, later could be earlier and the point at which earlier became later would be unknown.

If change occurs unobserved then it must accord with a series or order of things that is not temporal. McTaggart tells us that not only does 'time as it appears to us' not exist, but time-in-itself is non-temporal. Any sense of time is by definition phenomenal. Whatever time is, it must be, in some way, altered by the act of sensing. Real time, if it has any reality, cannot be the sensation of time. It must be the *a priori* existence of time-in-itself. But if time-in-itself is non-temporal, we are not, in our sensation of time, being aware of its true nature at all. Though its true nature has, in some way, stimulated our erroneous sensation of a temporal series.

McTaggart's universe is absolutely timeless. The idea of time, as we think of it, which is as temporal, only occurs in the misperception processes of our sensations. To follow McTaggart, we have to abandon objective reality — the seemingly obvious fact that things change — and come to terms with the idea that time, in truth, is non-temporal.

'Now' is always in the past. No matter what we consider as 'now', we will always have that consideration some time after the impression of nowness that we suppose relates to the object of our perception. In other words, whatever we think we see now, is in fact now (at this moment) in the past. Once we come to realise that it is in the past, it has moved deeper into the past.

There is no way around this. In a temporal world we cannot have a 'now' experience of an object. Even having a 'now' experience of ourselves presents difficulties. This would require that wherever our 'self' is it would need to experience nowness directly as part of the realisation of having that experience. Whether this is possible will be dictated by how we define the nature of our 'self'. Whether it is in any way confined to our body and if so whether it is confined to a particular part of our body. If it is not confined to our body it may be having 'now' experiences. However, our body (and consequently our body-mind combination) will always be late in realising it. If our self is integral with our body then we need to consider whether it is in all parts of our body or confined to an association with our brain. Its location, however, would do little to escape the problem, though there would be slight differences in the time taken for signal transmission if our self was in our big toe as opposed to our brain.

A selfhood that relates to a material body will always be 'behind times'. This can be illustrated by increasing the scale. Imagine a huge self, say one light year across, with arms and legs and a mind at its 'centre' associated with its brain. This mind is identified with the brain and relies upon nervous impulses to arrive at the brain to have any realisation of objects around it. The mind may have intimate relations with its own thoughts.

However, the thoughts or realisations it has about anything beyond self-reflection (of that strictly intimate sort in which the self directly reflects on itself), need to be brought in using the brain and its associated nervous system. This means that if it has a sensation, say a pain in its little finger, the cause of the pain will have occurred at least half a light year ago. Though this assumes nervous impulses travelling at the speed of light which of course they do not. Even so, if it responds to the pain by sending a message to move, then by the time that movement occurs, the need for the action will have passed.

If we are minds associated with matter then we are in just this situation. Though our experience will happen at the same time as the experience the (possibly objective) cause of that experience will be past. Possibly only a spiritual self that exists in a non-material universe can have 'now' experiences, where the experience happens at the same moment as the cause of the experience. Even this is doubtful if there is a relationship between the order of experience and the order of events. Though even if there is a relation its direction is not certain; the order of experience may be going the other way to the order of events. If this were the case (and we knew it), although we would have to reconcile how we experienced events 'before' they were caused, there would be moments (when the relevant ordering relations crossed over so that parts of each coincided) when both causes and experiences happened in the same moment. Though this is not born out by experience this does not mean that our experience is correct or what has been our experience so far will be our experience in the future.

In our everyday lives however, we would commonly differentiate between the causes of what we experience and what our perception tells us we are experiencing. What our experience tells us we are experiencing is thus infallible evidence of the 'not now' experience. This may be an illusion but it is an illusion we are prepared to live with. Based upon our previous experience, it is most unlikely that the illusory gap between the content of our consciousness and the cause of that content is meaningless. There is, after all, continual reinforcement to the contrary. However, if the causes of that experience disappeared during the milliseconds it took for us to be conscious of the experience, we would find ourselves suffering the consequences of the true pastness of the now just before we became aware of it.

It is perfectly reasonable to seek to replace the time-series. Indeed, we may feel that general confusion over the nature of time compels us to do just this. Much of our appreciation of temporality is a socially produced one. It may be that time is only a 'culture's perception of time'.[5] But it is not essential for survival. It is possible to live in a de-tensed world and still survive. There is much anthropological evidence for this.[6] Our view of time can take on one form when we are children[7] and another when we become old. Memory combines with anticipation to form some sense of us 'passing' or 'changing'. But this is dubious for if we look closely at our sensory images of the world we may feel uncommitted about believing in our own history anyway. Most things that we think of as 'real' have the common property of repeatability or sameness. This does not seem so certain with time because, even though we can impose upon it a strict metric, our experience of the passage of time alters. For example, time appears to go faster as we age or when we are well occupied. The only thing that militates against this is our reference to the established *Q* Series. If we accept the *A* series as a reasonable (though rule of thumb) description of the phases of change in time, matters only get worse, for at every turn it conspires to confuse. Accepting the *A* series is at the same time accepting never-ending contradiction. *A*-time seems inescapable and yet it is unexplainable, it seems real and yet it will not yield to any logical test. As Geach says, 'if McTaggart's alleged contradiction is bogus, we need not consider whether the attempt to remove it would generate a vicious regress; if it is not bogus, it suffices to show that *A*-characteristics are not objective'.[8] In the face of the *A* series we are easily made to feel like St Augustine — we can only understand it if we are not asked to explain it.[9]

That everything always seems to be in time does not disprove the idea that nothing is in fact in time. It may appear to be a conflict, but this is only superficial. If what appears to us as time is 'to all intents and purposes' like what we consider time to be, but 'really' is not time at all, we are presented with no additional problems as far as experience is concerned. Nor does our experience of time (erroneous or otherwise) affect the real nature of time. This reality, which is not time at all though we somehow perceive as time, is not affected by this superficial error. It is the same as Einstein's conclusion that although the real nature of time can only be found in the exactness of the Special Theory of Relativity, for everyday purposes the normal concept of relativity is acceptable. Similarly, our error of thought lies on the foundation of what it is that works for us. Our everyday life does not need a precision tool like the Special Theory to allow us to dwell in or deal with it. It is only if we want to view the universe as it really is we must view it in a different way. In the same way, if we want to view time as it really is then we must recognise that it is not temporal and cannot be described using everyday temporal terms. It is all very well talking truthfully about the nature of something but if our truthfulness is based on a misconception then we should be concerned about whether what we are talking about truthfully is in any way true.

That McTaggart's argument and our general puzzlement appear to show that the *A* series is anything but sensible (in that it does not truthfully reflect anything which is true) can lead us to look for an alternative. I think we are no less mistaken in looking for an alternative to temporality, than we are in trying to find a way around the compelling argument for the moon being made of cheese. As Dunne says, 'Where all is fog a blind man with a stick is not entirely at a disadvantage'.[10]

McTaggart accepts the pointlessness of a defence of the *A* series and sets out to identify the true nature of what presents itself so confusingly as time. Such a view is, of course, not the sole preserve of the metaphysician. Minkowski recognises the same inherent fallacy in the idea of an objective view of time and throws out temporality as an independent reality. He rejects the Newtonian absolute world and replaces it with a four-dimensional world, within which different 'projections' of space and time are experienced by different observers.[11] The boldness of approach is not dissimilar to McTaggart's. He rejects time as we experience it and the terms we use to describe it and replaces it with a system which apparently has little or no relation to all that we think of as connected to or arising from time.

The *B* series relies for its reality upon the *A* series. If the *A* series does not exist, we must assume that the *B* series is either something other than it appears to be or is simply an invention of human consciousness. It is probably a mixture of the two. McTaggart says we need not think that everything we experience in time is illusory, that 'whenever we have an illusory experience of a time-series, we are observing a real series, and that all that is illusory is the experience that it is a time-series'.[12] It is possible to imagine a *B*-world not reliant on consciousness, where ordering of before and after only had relevance to physics. However, *B*-relations in an unperceived world have a greatly reduced meaning. Although the *A* and *B* series are not temporal and are possibly erroneous perceptions, McTaggart believes that they can be indicators of reality. He thinks that what really happens in our experience is that we mistakenly apply the apparent terms and relations of the *A* and *B* series to the real *C* series. In our mistaken application we are therefore experiencing something temporal even though there is nothing really temporal for that experience to be based on. Even in our misperception, as McTaggart says, 'Through the deceptive form of time, we shall grasp some of the true relations of what really exists'.[13] What he means is that although our perception of time does not allow us to know what time really is, that perception, because it is caused by the true nature of time, nevertheless somehow reflects what it really is. In other words, every (timeless) time I perceive a brick I think I perceive a clock, but I cannot perceive a clock without perceiving a brick so there must be some 'clockness' in the nature of the brick even though it does not in any way resemble or work like a clock.

According to McTaggart, the *C* series is real and the *A* and *B* series are only apparent. The *C* and *A* series he considers primary and the *B* series secondary. For us to have the experience of time, 'The real *C* series and the appearance of the *A* series must be given, separately and independently'.[14] Both series are essential to time and they cannot be derived from each other though a real *C* series of permanent relations will appear to form an *A* series. The *B* series, however, in its relation from earlier to later, will appear to derive from the *A* series and if we accept the *C* series then the *B* series will appear to derive from that as well.

If we perceive events as appearing simultaneous, we are in fact having a true perception, because these events do occupy the same position in a series. However, this is the non-temporal *C* series so their 'simultaneity' is more a 'similarity of placement'. Again, if we perceive events to be in a time series, we will be truly perceiving them in a series. However, the truth of our perception will lie, not in a time series, but the non-temporal *C* series. According to McTaggart, 'the *C* series will include as terms everything which appears to us as an event in time, and the *C* series will contain the realities in the same order as the events are ranged in by the relations of earlier and later'.[15] But this does not mean that the *C* series is identical to the *B* series. The terms of the *B* series are events and the terms of the *C* series are not. The uniting relations of the *B* series are the relations of earlier and later and this is not so with the *C* series.

McTaggart does not deny that what we think of as time in fact exists, nor that it has as part of it many features of what we think of as time. What he says is that this thing that we think of as time, though a series, is not a time series (that is the series of asymmetrical and transitive phases of change in the *A* series). Though what we are aware of is real but what happens to that awareness in the process of making it sensible to our minds, causes error. Because reality is in this way hidden, we build up a picture of reality based upon a mixture of misperception and misinterpretation.

We find it easy to believe in the straightness of the apparently bent spoon dipped in a glass of water, only because the straightness has been shown to us and we have felt the spoon with our own fingers. Though every time we look at it in the glass still it looks bent, we would laugh at anyone who believed it was not straight. It is the same with time. We can imagine that there is a sort of bent spoon, which we call time, but really it is not bent and so should not be called time. We cannot take it out of the glass to see it 'as it really is' and we cannot feel it with our hands, but its 'real' nature nevertheless can be described. The 'proof' of the spoon's straightness is available to us whereas the proof of the reality of atemporality is not. However, our confusion about time reinforces the feeling that it's 'bentness' may be corrupting its actual nature. Because the description of the atemporal reality behind time seems more fantastic than the apparently real, this does not mean that it is untrue. As Aristotle says, 'A likely impossibility is always preferable to an unconvincing possibility'.[16]

If we need weight to add to our feeling that all is not well with time, we need only try to analyse its elements to find ourselves immediately confused. To accept, as real, something that presents so much self-contradiction, and demands so much mental labour to try to answer those contradictory features, is perhaps philosophically perverse. On the other hand, to accept that such a high degree of self-contradiction suggests something, the coherence of which is suspect, and look for a replacement, may be philosophically enlightened.

The individuality of selves presents particular problems for temporality because selves in the atemporal world amass experience without any need for temporal ordering. Oakeley[17] thinks this central in McTaggart's argument against time. But this is, in some ways, the wrong way around. McTaggart's argument against time is only part of his description of the self and the universal system it forms. But we do not always need to view atemporality in this context. Certainly we need to appreciate that selfhood in the *C*-world is always atemporal but atemporality stands without having to accept the pluralistic spiritual selfhood that McTaggart proposes.

To pass beyond the processes of change to some sort of realisation of the eternal, the individual must deal with the illusion of change in every part of its experience. It is the illusion of the separateness of selves, and that time must be unreal (as Spinoza thinks), which is the difficulty. Oakeley thinks that what seems unique in McTaggart's system, 'is that whilst his reason for this position is a logical doctrine of the contradiction which arises in our propositions about anything which is subject to change, his pluralism of real selves is determined by a spiritual doctrine of experience'.[18]

Oakeley believes that McTaggart's idea of the self is, in fact, incompatible with the unreality of time. She thinks the inclusion series does not seem capable of linking the temporal and the eternal unless, as she says, 'the temporal order really lies half concealed behind the relation of inclusion'.[19] This seems a misplaced concern. It is not appropriate to describe the temporal order in this way. The *C* series, the relation of inclusion, does not 'conceal' the temporal order. It is more that the temporal order is a misperception of the inclusion series. It therefore has no reality other than at a perceptual level and McTaggart urges us to accept a reality beyond the scope of our perception. If we focus on a reality beyond perception then the temporal order is no more than an incorrect (though unavoidable) mix-up in the way we see things. In some ways it is like watching a film that gives us a different view of temporal ordering; indeed, so different that it is atemporal ordering. The fact that we are witnessing the ordering in our own time-order does not alter the fact that what we are witnessing is atemporal. What McTaggart encourages us to do is suspect the reality of our own framework and replace it with the image of the atemporal world. It is interesting that even if the film image is portraying time we can have reason to suspect its temporality. Currie argues that cinematic images do not have tense in the *A* series. His view is that although such images are anachronous and although anachrony is explainable in terms of temporal *B*-concepts its direction is not.[20]

In part, we get this mixed-up view about time from the nature of our individuality and in this we are condemned (at least for a long time). However, the nature of this individuality is progressively altered as we come to know other selves in the inclusion series. Only when fully we know all other selves, will we be able to see the truth of the reality of the *C* series.

But are we deceiving ourselves by looking for this reality? Perhaps the ontological quest is unreasonable. Maybe, as Dummett thinks, we must abandon our deep-seated belief that there must be a complete description of reality. Dummett defends McTaggart against three commonly used lines of attack: the predicates of the *A* series, events and McTaggart's ignorance of token-reflexives.[21] Dummett argues against those who say that it is not just the predicates 'past', 'present' and 'future' which are in question but also 'was past', 'will be future' and so on and that the addition of these removes the contradiction. Dummett thinks this a meaningless hierarchy that does nothing to remove the contradiction, for no matter how many levels are added, the first level predicates of 'past', 'present' and 'future' remain. Thus, any second level past, present or future is in the past, present or future of the first level, and any third level past, present or future is in the past, present or future of the second level which is in the past, present or future of the first level and so on.[22]

As for events, Dummett thinks there may be no need to use, in our language, any expressions denoting events. It may be satisfactory to use only names of, or generalizations about, the perceptions in question. For example, every event may consist of something happening to an 'object'. Time involves change and no matter how we refer to events (here as objects) they will have different predicates referring to them at different times. By setting aside temporal words the result will reveal incompatible predicates just the same. For example, 'the grass that was green is now brown', becomes the two predicates 'green' and 'brown' both of which apply to the grass, but without any means of explaining the incompatibility of the two characteristics.

It is often said that McTaggart's argument is a sophism based upon ignoring token-reflexive expressions. McTaggart is accused of arguing about time without setting aside temporal language, getting involved in verbs and tenses, or emphasising that the language he naturally uses, actually relates to the *C* series and the implicit qualitative improvement that is its nature.

A token-reflexive expression is one that renders a sentence capable of different truth values according to the circumstances under, or in which, it is uttered. For example, using such expressions as 'I', 'here' or 'now' will have different effects on the truth of a statement according to who said it, where they said it, when it was said, to whom it was said, what gestures accompanied it and so on. Based on this, if we take a token-reflexive expression to apply to an existent if there are some circumstances under which it can apply, then there will also be another set of circumstances under which another token-reflexive expression could equally apply. If then we say that token-reflexive statements must be incompatible if there is no set of circumstances in which both of them can be truly asserted of the existent, it would appear that McTaggart's contradiction is not a contradiction at all. Dummett points out that were this really the case, 'then only stupidity could explain McTaggart's failure to use a quite analogous argument to show the unreality of space and the unreality of personality'.[23] Such a view would show every place to be 'here' and 'there' and 'near' and 'far' and so on and would make 'me', 'you' and 'him', 'her' and 'her', 'me' and so on. McTaggart did not use any such argument and often draws the real distinction between space and time. To show this we can define McTaggart's argument as having two parts in this way. Part 1. There would be no time if there were no facts of kind (a) on the grounds that time involves change but change can only occur if there are facts of kind (a). Part 2. Facts of kind (a) involve a contradiction. Dummett says that Part 2 therefore depends upon Part 1 (which is the heart of the argument) and it is because the analogy of space or personality cannot be applied to Part 1 that Part 2 has no force in these contexts.[24]

Dummett does not accept that it is possible to observe a four dimensional manifold of space and time, because the conscious observer needs token-reflexive expressions to explain what he is observing. According to Dummett part of what McTaggart is saying in Part 1 of his argument supports token-reflexivity because it cannot fully be described without some use of token-reflexive expressions. Indeed, it might be interpreted as supporting the reality of time because it shows that time cannot be reduced to anything else. On top of this, McTaggart's neglect of token-reflexives could appear to render Part 2 of the argument invalid.

Objects in space could be described by an imaginary being who was himself not in space, though according to McTaggart it is not possible to describe positions in time unless the describer is himself in time. Of course, it would be possible to stand outside time and describe it, in the same way that we can stand outside space and describe it, but to describe the temporal events as part of our experience, we must reserve a place for ourselves in that series of temporal events. An observer who is observing events in time cannot choose the order in which he will observe those events. They will occur only in the order in which they are witnessed. It may be that the witness experiences events in the wrong order. For example, the witness may say that 'this' and then 'that' happened, when in fact it was 'that' and then 'this' that happened. The order of experience may not be the experience of order. There is also the question of whether the witness's thought of the experience is the same as experience of the thought of the experience. In all cases it remains that it is an order (temporal or not) which is the object of the experience. The observer will always be observing from the temporal position of 'now' and if he were put 'into' the time sequence we would need another observer. Even so, this second observer would make an equally fruitless attempt at observing, and so would a third and so on. Only if time were represented as a three-dimensional space-frame could the events be observed in the same way that space could be observed. But even then, this would only be a particular slice (though maybe a large slice) of time, and any observation made would be subsequent to the last one.

McTaggart's view is that in order for something to be real it must be capable of complete description. Because it is not possible to construct a complete description of temporal events (as just pointed out) then time cannot be judged to be real. Dummett supports this, himself believing that observer-independent complete description is necessary to reveal reality. He is suspicious, however, (as McTaggart was of evil) that the illusion of time itself, presents us with a view of something that is in time. Even if it is an illusion it is a temporal illusion.[25] Nevertheless, he believes McTaggart shows that we must abandon our deep-seated belief that there must be a complete description of reality, if we are to hold with our belief in time. McTaggart, of course, offers something he feels can be completely described and is atemporal in the *C* series.

But we should be wary of throwing out the notion of a complete description of reality because of its apparent inaccessibility. King-Farlow defends McTaggart's argument against time as part of a greater scheme, the *Search for Something Deeper which Underlies our Sense of Time's Passage*, as he calls it.[26] *The Search* is another way of describing the ontological quest, the mystic cause or the fuller purpose. It was born in the fifth and sixth centuries BC with the Indian and Greek philosophers, given vigour in the west by Leibniz and Kant, turned on its head by Bergson (Bergson put the concrete above the abstract and gave direction precedence over scientific concepts) and more recently undertaken under the guardianship of physics and mathematics by Reichenbach, Grünbaum and Gödel. King-Farlow is unusual among McTaggart's critics in that he refers much more widely to a range of McTaggart's texts than do any of the others.[27] He feels that too many critics have aimed too much negative criticism with reference to too little of his work. He attempts to fill, what he calls the 'positive leg' of McTaggart's trousers, for the sake of history and for the sake of McTaggart's assertions about our experience of what we call time. He thinks that McTaggart is a major contributor to the *Search for Something Deeper*.

King-Farlow says, 'McTaggart, Reichenbach and Grünbaum are alike in showing a liberality towards events ... for them there is no sin in accepting as fully objective some series of events forming cosmic history'.[28] He also thinks that they are together in rejecting, 'Our vulgar "*A* Series" notions of time ... concepts of Absolute Becoming and Time's (Becoming-directed) Arrow'.[29] He reminds us that McTaggart thought much about the relation between time's arrow and its relation to the historical stages that represent the Absolute. Also that by the time the 2 volumes of *The Nature of Existence* appeared in the 1920's, some Einsteinian philosophers had decided that causality was at the root of the ordering of events, though others, Reichenbach and Grünbaum, fixed on something different as an answer to our psychological problems over time (an order in the overall increase in entropy).

King-Farlow recommends that some mental energy should be given to trying to, 'illuminate certain fascinating likenesses between McTaggart and the Entropists, between a paradigmatic metaphysician and those who affect to despise any creator of an abstract metaphysical system even while they pore over an exceedingly abstract system of Space-Time geometry'. King-

Farlow thinks Gale, in his own work and in his selection of the work of other critics, 'creates quite unnecessary befuddlement as to how certain of the Cambridge metaphysicians' tactics could ever make any kind of strategic sense'.[30] Further, King-Farlow reminds us, that in Gale's *The Philosophy of Time*[31] and *The Language of Time*, not only do we only get fragments of the text from *The Nature of Existence*, but neither McTaggart's 'The Unreality of Time' nor 'The Relation of Time and Eternity' are mentioned. Nor, as King-Farlow points out, does Gale (and subsequently so few others) properly explain the *C* series, which is the crucial non-temporal reality of substances that replaces the debunked *A* and *B* series. Gale refers to the *C* series only three times in *The Language of Time*,[32] and then only after introducing it as 'a series whose generating role is non-temporal, such as the series of integers'.[33] For McTaggart, the *C* series is much more than this. It is what represents the notion of increasing representational adequacy.

I think King-Farlow correctly recognises that the *A*, *B* and *C* series have to be viewed together. He points out that underlying the illusory *A* series, which seems implicitly to have a sense of direction, is an objective sense of direction of events which, although epistemologically less accessible, nevertheless, can be realised. Although, on the face of it, this objective sense seems more obscure it can still reflect the ultimate reality.

King-Farlow classifies *A*-theorists as those who believe that the concept of Absolute Becoming (no matter how deeply imbedded in ordinary language or common sense) is essentially worthless in any philosophy of time.[34] He categorizes three up-dated versions of McTaggart's idea. 1. Ontologically, let the Ultimate Reality be a Space-Time manifold. Let the 'time' element have taken out of it any concept of 'vulgar time' or 'absolute becoming'. King-Farlow notes that this idea should be compared to that presented by Williams.[35] A view he calls 'a (would-be) *B*-Theorist's [Russell is a *B*-theorist, Gale and Broad are *A*-theorists] classic on "the four-dimensional fabric of juxtaposed actualities"'.[36] Grünbaum reinforces this.[37] 2. Phenomenologically, let every new 'revolution' in science be understood to offer a more adequate representation of 'What There [Tenselessly] Is'. 3. Scientifically, let the emergence of laws such as entropy or evolution consolidate the wisdom of the substitution policy in 2. This categorization fits into McTaggart's idea of increasing representational adequacy.

From this, King-Farlow helps us to re-describe McTaggart's positions on the unreality of time by providing five pointers. 1. It is false to say time is unreal because we see time as a series of changes. What we misperceive as time (which is a true misperception) is, in fact, a changeless real substance. 2. It is false to say time is unreal because we think it is an illusion. Our perception of what we think of as time gives us many clues to the ultimate reality and as indications they are not illusory. 3. It is false to say time is unreal because we see temporal directionality as a kind of linear ordering. Such ordering cannot be relativistically so to some observers and not to others. The true nature of temporal ordering would make the latest events more representative of the Absolute than earlier events. 4. It is true to say time is unreal if what we are saying is that real ordering of events (in the *C* series) leads from less adequate to more adequate representations of reality, as opposed to our vulgar belief in the self-evident objective *A* series. 5. It is true that time is unreal if, by *a priori* analysis, the *A* series and the concept of Absolute becoming can be shown to create contradictions (or a vicious infinite regress).

King-Farlow thinks McTaggart is easily mishandled because of a misunderstanding of his overall strategic intention, what he calls 'his curious weakness regarding the *C* series'.[38] But King-Farlow tells us that McTaggart's real offering is in his metaphysical strategy and not in his providing fighting ground for the criticisms of sceptical tacticians. He also thinks that the establishment of the consistency between his two articles 'The Unreality of Time' and 'The Relation of Time and Eternity' provide great reward to the philosopher.[39]

There is, in the work of King-Farlow (and Welby and Shea), strong reason to think that McTaggart's denunciation of time should not be dealt with as if it were an end-in-itself. We need to withdraw from the argument over the adequacy or otherwise of the details of McTaggart's argument and its ability to successfully denounce time and instead move towards a fuller analysis of McTaggart's atemporal replacement. The idea of matching McTaggart's view, first with his fuller concept of the *C* series and then with the modern scientific view of space-time, is a fruitful though relatively unexplored avenue. It has recently been looked at by Yourgrau[40] and I will look at it in § 5.4.

5.3 Reality of becoming plus consciousness

This section, 'reality of becoming plus consciousness', draws a negative conclusion that inadvertently gives McTaggart some support. This position is commonly called the 'dynamic' view and it is the one McTaggart challenges. The discussion here highlights the need for there to be 'something' in which 'something happens' and this throws us back into the confusing nature of the *A* series.

The notion of temporal becoming rests on a belief in the *A* series. There has to be something 'out of which' something can 'become'. This does not presuppose a future that holds all the events that eventually become. It presupposes that at the moment of becoming, the events that become, emerge from somewhere or something. Becoming can be viewed much like the flow of the river of time. That is, either as us moving into something that causes events to become in our experience, or as events moving to us and causing in us the experience of becoming. It can also be a 'being' into which it becomes. Whichever, it remains that 'becoming' is the consciousness of the next experience in time. Becoming is the adding of facts to the existing totality of facts and is a belief deeply embedded in the human psychology. As Wittgenstein says, 'The world is the totality of facts, not of things'.[41]

For temporal becoming to make sense, there has to be something that is conscious of that becoming. Without consciousness, it is impossible that anything could be understood as 'becoming'; rather misusing Wittgenstein, 'Whereof one cannot speak, thereof one must be silent'.[42] Without consciousness the adding of new facts may well go on, but without the conscious appreciation of those additions, there are no *A*-concept and no 'becoming'.

If considered separately, there is a basic difference between the three dimensions of space and the one dimension of time. Physical events presuppose the spatial dimension, while mental events presuppose the temporal dimension. In addition, whereas we can speak of the 'passage of time', we cannot speak of the 'passage of space'. The metaphor of the flow of time, relies upon our mental appreciation of the asymmetrical nature of time that has no spatial presence.

We can think of the flow of time either as events flowing 'past' us — coming from the future through the present and into the past — or as us moving 'through' it from the past through the present and into the future. Which one we choose is a matter of how we view the relative nature of our self and the 'flow'; that is, whether we are standing still in a river or are moving in water that is standing still. Whichever we choose, we are placing our self central to the process. The idea of the flow of time is one that dominates our common sense thinking and fits well with the metric of time. But there is another alternative which still gives us a centre position. The notion of temporal becoming proposes that events progressively come into being at the present time. The concern here is, not so much where those events come from or go to (though they must 'come from' and 'go to' somewhere) but that they become known by our consciousness. Because temporal becoming is instantiated to our present, we must consider whether the events that become are objective, or whether the phenomena are entirely subjective.

Bergson[43] thinks that physics identifies time as an *A* series of strict asymmetric order, *le temps*. This equates with the Newtonian idea of the 'equable flow of time'. Bergson believes that our conscious appreciation of time is different from this. What happens to our consciousness, is an awareness of a continuous unfolding of events into the future, the *élan vital*. The *élan vital* is conscious of the real nature of what is temporally going on, *la durée*. There is much in common with this and McTaggart's separation of the apparent but erroneous *A* and *B* series, from the non-apparent but true, *C* series. This is a theme underlying the thinking of others including Husserl,[44] Merleau-Ponty,[45] and Heidegger.[46]

As already noted, Merleau-Ponty thinks that events can be in no way objective. As such, time cannot be a real process but 'arises from *my* perception of things'.[47] This relation to things is a mental appreciation that progressively pushes out into the future. Though, if we take consciousness out of the equation by removing all living things, still we tend to think that something objectively temporal would be going on. More simply, if there was a time before living things, then something temporal must have happened for us to be where we are now? Merleau-Ponty thinks that all things in the past can only be made sense of by a conscious appreciation of them now. As he says, 'Laplace's nebula is not behind us, at our remote beginnings, but in front of us in the cultural world'.[48] Merleau-Ponty therefore demands that the present be the only place where anything can be known, and that knowing is in the domain of the present perceiver as verificationist. Common time, which is a mutual understanding of the present or past events, is found in the subjective world by what Merleau-Ponty sees as the 'living present'. The living present is the bodily existence we take part in and by which we share common understanding and knowledge with others. This form of temporal becoming is therefore entirely anthropocentric *A*-thinking.

The place of consciousness in temporal becoming is also found in *B*-thinking, in that all that is necessary for our appreciation of time is to be found in the succession from before to after. A contemporary expression of this *B*-theory view is found in Mellor[49] though, as we have already seen, its foundations rest with Russell. Ayer also argues this case, notably in his denunciation of Broad's *A* series view.[50]

Mellor thinks that the *A* series cannot simply be reduced to the *B* series, but that tenseless terms can be used to express truth conditions for tensed sentences. In Mellor's thinking there is no place for tensed facts that only make tensed sentences and judgements true or false. The truth or falsity of such tensed sentences and judgement tokens is already determined by tenseless facts. He says, 'Provided a token of "*e* is past" is later than *e*, it is true. Nothing else about *e* and it matters a jot'.[51] Any statement about the past, for example, '*e* is past', is true if, and only if, the utterance of the statement is later than *e*. The truth conditions of the statement '*e* is past' can be found exclusively in relation to the terms of facts in the *B* series. Although tensed expressions may or may not be indispensable, to fulfil any truth conditions we need never resort to tensed facts. From this, he concludes that the truth conditions for tensed statements are either tenseless or contradictory.

I think that neither the *A* series view nor the *B* series view (as illustrated by Bergson, Merleau-Ponty and Mellor) is sufficient in itself to deny the objectivity of time. It remains that something triggers the conscious response in us that something is happening that seems like time. Unless we step aside, like McTaggart, and deny the reality of time, we will always be stuck with the possibility of some sort of objectivity. Other attempts at a solution to this get us no further, indeed, they can make matters worse. Schlesinger tries to escape the *A* series paradox providing a transient solution within the *A* series by relativizing statements about present worlds in which they are true.[52] Biglow further develops Schlesinger's view.[53] This argument is taken up by Buller and Foster who think that Schlesinger's relativization reduces *A*-terms to *B*-terms and therefore introduces a parallel paradox.[54]

A *B*-world is not adequate for a meaningful conception of time. To make sense of *B*-relations, we use as common ground, our *A*-thinking. Without this, an exclusively *B*-world may not appear temporal at all. A world without conscious beings may not have *A*-terms ascribed to it, but even so, there would be the possibility of such conscious ascription. It is time itself that provides this possibility and the possibility is affected by our being. It is the conscious sense of becoming that picks up the hint of objective temporality.

We can interpret the consciousness of time in different ways, but basically, it is whatever links our appreciation of temporality with what is actually happening temporally. The acceptance of this has a long lineage. Particulars can only become wholes, in Kant's thinking, by a conscious 'gathering together' or by 'touching' the forms of space and time, as such the ego 'sees' the process.[55] Plato also interprets thinking as a kind of 'seeing'. The same process of mind leads, through Hegel's synthesis towards the absolute, to McTaggart's perception of selves in the *C* series. Aristotle rejects this kind of thinking and proposes a mind that can actually grasp the essential being of a thing.[56] Rosen discusses whether we 'see' or 'touch' reality and the link this has to our thinking (in particular respect to Aristotle).[57]

Temporal becoming relies upon a belief in the future and an acceptance that, at the very least, 'something was about to happen and did', or even, 'something is going to happen (and then does)'. This can be taken further by modal thinking which can be concerned with what might, must or could not be the case.[58] On this view, there may be only one actual present or past, but there is no limit to the number of possible futures. In such possible worlds, all of which have variable truth values, one (mine) is known, the others, though unknown, still may be true. This means that no statement about my future can be reliable for other possible futures. But whatever the nature of the future is while it is future, when it becomes present it must be whatever case it is. The question about what will be the nature of that happening is irrelevant to the happening when it occurs. The riddle of whether there will or will not be a sea battle tomorrow cannot concern us when the sea battle does or does not occur. But to believe in the notion of temporal becoming, accepting that there will (without question) be a tomorrow for something to either happen or not happen in, is all that counts. Without a future there can be no more becoming. This way of thinking is, according to Lucas, no longer *esse est percipi*, to be is to be perceived (where Berkeley does not concern himself with any temporal future), but instead *genesthai esti horathēnai*, to happen is to be perceived (an expression he uses to describe Schrödinger's extension of Berkeley's thinking). Lucas says, 'Schrödinger has out-Berkeleyed Berkeley, for whereas Berkeley was talking only of things existing, and saying that for something to exist was for it to be perceived: Schrödinger extends this to events taking place'.[59] Temporal becoming requires 'happening' and 'happening' requires future and present. 'Happening' is the change of state in the thing, it is either events that are changing or things that are changing. In either case, nothing can change unless there is an 'opportunity' or 'possibility' for that change to occur in and that requires a present into which a future event can become.

5.4 A four-dimensional universe with no becoming

This section, 'a four-dimensional universe with no becoming', surprisingly puts McTaggart back in the centre of a scientifically valid position that, at first sight, we might think hostile to his idealist argument. Here I show that idealist thinking, far from being superseded by modern scientific views, combines meaningfully with them. Taking this further, I explain how I think that McTaggart's atemporal reality can make a direct contribution to four-dimensional thinking.

From what has been said so far, the chances of finding temporal reality in the *A* series seem remote. Serial asymmetry of temporal order seems improbable when we analyse each phase of change separately and it is only a short step from this confusion to an acceptable notion that the *A* series is self-contradictory. There is, however, a third alternative and this means looking at the scientifically validated theory of relativity. Here, though we may find the same negative conclusion we may also discover something positive. For example, there is some support for the reality of the present. Sklar[60] (developing an argument first put forward by Rietdijk[61] and Putnam[62]) argues for the reality of any observer's 'now', whether it is in the future of another observer or the past of another. In this way future events have as much determinate reality as any other events, past or present. Though there is also a strong belief that the four-dimensional framework eradicates any further considerations on time as a thing-in-itself. Putnam denies that there is any more philosophical enquiry to be done on the matter of time. He thinks scientific investigation of the four-dimensional framework is all that is left. Teichmann[63] (in trying to follow Shoemaker[64] on the post-verificationist view of the idea of time without change) concludes, like Putnam, that logical and linguistic tools are, by themselves, incapable of determining the truth value of the statement. His view is that 'Time without change either has elapsed, or will elapse' and that such things are best decided by science.[65]

However, we should feel philosophically unconvinced by this view of the over-arching 'rightness' of science. It does not, after all, answer any of our everyday confusions about time by taking the high-ground with such authority. We can be more suspicious of science when its view leads to metaphysical conclusions. If there is no privileged observer, and there cannot be in space-time, then there can be no absolute simultaneity. If there is no simultaneity then past, present and future are equally real or unreal. The question of the nature of what it is that is 'real' or 'unreal' then becomes the problem. It seems unlikely, if we have given up linear, temporal asymmetry to find ourselves here, that we should re-invoke it to determine the nature of reality.

Sklar thinks that there is some metaphysical mileage in taking this 'step back'.[66] If we invoke an observer at a time, then everything in his future or past must be unreal. Though, as Putnam thinks, events in the past can be attributed a truth-value that accords them some degree of reality. Indeed, there may be events that will be in my past at some future time but never be real to me at any present time.[67] But everything outside the light-cones of the observer's future and past are in the region of the absolutely simultaneous or absolutely present. To deny this categoric reality seems obscure. After all, it is from here that the light-cones both 'gather up' and 'dispel' the observer's causal future and past. In addition, two independent observers will attribute reality and irreality to the same space-time objects. If the observer's present is real, the only reality that anything outside the observer's present can have is as a reference-value to the present.

The past and future seem unreal due to the epistemic 'distance' that they have from the reality of present. But this distance is meaningless: either we have epistemic 'contact' with events or those events are epistemically unreal. Epistemic contact can only occur in the present (of the observer), so everything in the observer's future and past light-cones and everything outside them, is unreal.

Sklar suggests that there is a way of blending our idea of space and time, so that the epistemically distant object in space, is no less real than the epistemically distant object in time. He thinks that from a pre-relativistic standpoint, there is a tendency to affirm the reality of the temporally distant, while denying it to the spatially distant.[68] In this way (he suggests), a natural symmetry is re-introduced based upon the conceptual structure of space-time. This allows us to shift away from the verificationist solipsism of the present moment.

Verificationist thinking is inherently bound up with the *A* series. In the 'indistinguishable space-time' of General Relativity, no observer can tell which space-time he inhabits, only a 'super observer' could determine this.[69] Though Einstein moved away from the verificationist view, traditionally it remains inexorably bound to relativity theory and we would not expect to find McTaggart's idealism here. However, the place of consciousness is important in his theory, and the four-dimensional framework has remarkable similarity to the *C* series. What is common between the *C* series and the four-dimensional framework is that neither of them is strictly temporal. To pursue this requires a certain 'spatializing' of time.

Welby[70] points out the reliance that our terminology of time places upon motion and space. She highlights the incongruity of speaking of 'series of positions running from the far past through the near past to the present, and then from the present to the near future and the far future'.[71] Yet it is these sorts of terms we naturally rely on, both in our expression of temporal things, and our instinctive concepts of temporal things. Because of this we too easily ignore distinctions of the words 'past' from 'passed', 'present' from 'presented (here in this place)' and 'future' from 'yet to come'. Welby says, 'When we admit that time involves change, may we not rather suggest that change shares with Motion and its implied Space wherein to move, the parentage of Time, as movement through the space of experience in an irreversible direction'.[72] Welby accepts that the relations in the *B* series are permanent, because earlier events are in a 'marked off *area* of experience' and an event 'never ceases to have a *place*'.[73] Unless change is derived from, or dependent upon, time (which seems unlikely) there cannot be a 'timeless change'. In order for change to change it needs both 'room' and 'time' in which the changing can occur. Change can only be a change in the characteristics of events as they are presented in the *A* series. Neither character nor presence is temporal, character is qualitative and presence is spatial. Events in this way are not temporal for they 'take place'. The error of spatializing time runs deep and it is no surprise that we find it difficult to explain past, present and future. Nor should we be surprised if we find ourselves thinking that whenever we judge anything to be in time, we are in error. We are in error because the events we are trying to recognise in a temporal sense are, in fact, dependent upon motion and space.

Welby places past and future on a sort of drawing board. We are presently closed to the future, but it is available by using science and disciplined thought. This view regards time as a 'room for experience'.[74] Our ability to predict is based upon our true knowledge and she thinks McTaggart is leading us astray in portraying time as an original category, as opposed to a deviation or secondary product of motion and space. The opposite view to Welby can be found in Markosian.[75] Markosian (though confusingly if put in the context of McTaggart as he introduces spatial terms he calls, *C*-properties and *D*-relations that are not to do with the *C* or *D* series) defends the idea of the pure passage of time (as an absolute) which is relativized only for convenience against things like clocks or the passage of the sun.

Shea,[76] in a way, begins again the idea that Welby introduced in 1909. He explores the link with McTaggart's argument to space-time. He defends McTaggart against setting aside temporal language and not getting involved in verbs. It is from this point that our argument for the connection between the *C* series and space-time proceeds.

Shea explores the 'surprising bond' between McTaggart's search for reality beyond time and the writings of Reichenbach[77] and Grünbaum.[78] He refers to *determining correspondence* as a means of introducing an altogether more interesting idea. Shea highlights how McTaggart relates *determining correspondence* to 'clearness of perception'[79] and that this clearness of perception will oscillate in time, sometimes appearing more clear, sometimes less clear, but always within the context of a trend that is moving towards increased clearness.[80] The similarity to this and the second law of thermodynamics is unmistakable. By this law, a state of entropy may increase or decrease along the temporal axis of any physical sub-system, but always in the whole system of the universe. The increase in entropy is one-directional along the same temporal axis. Faith in this law (because after all that is all it is) makes the calculations and representations of time in Minkowskian space-time much easier. Though it is, after all, a law that has to bear much evidence to the contrary. As Shea points out, 'entropy actually decreases in the *Lebenswelt* of intelligent organisms'.[81]

Shea thinks that there is some resemblance of 'partial event-substances and the Absolute'[82] which can be identified in *determining correspondence* and clearness of perception. In the *C*-world good things have preponderance over all things, lovingness and happiness will have preponderance over all wickedness and misery. That despite some retrograde oscillation in any, or all, of these things the overall trend of the real series (the *C* series) is towards greater clarity, goodness and lovingness. In the space-time world the irrevocable movement towards entropy is itself a movement towards 'clearness' and 'purity'.

The cosmologies of McTaggart, Reichenbach and Grünbaum all present theories with a sub-series or sub-system level, the actions within which may conflict with the 'real' or 'overall' system. In McTaggart's case, that our perception of time is erroneous but that via this misperception, we are getting a hint of the true nature of the true *C* series. They all show us that time's appearance can only ever (at best) partially resemble its real nature. Reichenbach assumes that increase in entropy defines the direction of time. In sub-systems (he calls them branch systems), any apparent decrease in entropy (he talks of 'anisotropy' and 'shuffling'[83]) is described in terms of an overall increase in those systems. This fits in with the apparently general increase of entropy in the whole universe.[84] However, Reichenbach admits that our view of time is so limited that 'it is very difficult to come to a conclusion concerning time as a whole', and that 'it is not necessary to assume that time has one direction throughout'.[85] Grünbaum accepts, like Reichenbach, the principle that the positive direction in time can be inferred from the apparent increase in entropy in branch systems. Though he is more cautious than Reichenbach, only assuming that the apparent increase in branch systems gives us reason to believe that the general trend is towards increased entropy.[86]

Shea points out that Moore sought to demolish 'negative' philosophies like Hume's and 'positive' philosophies like McTaggart's (when they offended against common sense), with two basic uses of *modus tollens*.[87]

That is that if I know it by common sense then there is no point in setting up an argument against it. Any argument purporting to be true, cannot imply the falsehood of other truths. Shea resorts to a similar principle. He says that the unidirectionality of time is so firmly entrenched in our minds as common sense, technological sense and scientific sense, that the idea of accepting sub-series or branch system non-unidirectional time is unthinkable unless the practical gains are overwhelming. If we do not consider a glimpse of reality a 'practical' gain then we should agree with him. If, however, reality is epistemologically important then Shea has it the wrong way around. McTaggart, Reichenbach and Grünbaum are saying that what we temporally apprehend is not what time is really like. Our experience of time utilises a branch system or sub-series, that although suggesting the real time series, does not strictly resemble it. If this is so then there is no point in replacing our practical (if slippery) grasp of the time series as we experience it with another version or interpretation of time as something temporal. We have to ditch time altogether if we are to find any real truth.

Yourgrau thinks that there is some connection between McTaggart's truth behind the unreality of time and the scientifically vouched-for space-time theory. In doing this he is highlighting McTaggart's ontological search for an ideal world that can only be understood from his fuller purpose. If we suppose (like Eddington) that relativity theory can lead us to an ontologically 'truer perspective'[88] and take on the idea (like Luce) that, 'Parmenides' sphere of reality [much like McTaggart's end-stage of the *C* series] seems Einsteinian rather than Euclidean',[89] we have a sound basis for comparison. If relativity theory can provide grounds for an ontological search then (even if only in that respect) it is similar to McTaggart's *Search for Something Deeper.*

McTaggart recognises that time is unreal and its appearance does not reflect reality. He believes that no one 'point in time' is available from which to extrapolate the *A* series as a real series. He understands the token reflexive nature of the *B* series, but denies its reality unless an *A* series exists. He denies the specious present and considers that our only possible, though erroneous, view of time is *sub specie temporis*. He concludes that the *C* series is real, *sub specie temporis* as time, but in truth it is a non-relational series that includes all apparent matter and apparent temporality. He finds that there are no fixed reference points and that time cannot be distinguished from space. He judges change, in the sense of actual change, to be unreal.

In a similar way (though of course by diametrically different methods and with very different motivations) Einstein recognises that Euclidean mathematics does not reflect reality, that the appearance of time does not match with any conventional alternate view that combines both space and time and that space and time are only fixed arbitrarily.[90]

This is an example of what Sklar calls the 'interdependence of science and philosophy'.[91] These two theories are attracted to each other, not by their opposite qualities, but by their similarities. A brief review of these will serve to demonstrate (if not in any way prove) the point.

1. Many Series. The *C* series involves as many time-series as there are perceivers or selves. In all pre-final stages the self is limited in the degree of perception available to it. Space-time provides each observer (or percipient) with a confining light-cone. This light-cone describes the limits of that observer's knowledge of the world.

2. Overlap. The *C*-theory tells us that there is an ultimate total unity of perception as each perception, by one self of another, passes on down a chain of perception. Space-time worlds overlap one against another so that one world is 'connected' to the most distant world.

3. Relative nowness. We have seen that McTaggart accepts that the specious present of any individual may overlap that of another. Each experience of time is a separate experience, there can never be a shared or common experience. Similarly, space-time cannot adapt to a common experience. The coordinate system x,y,z,t can only accord with a particular observer.

4. Time is unreal. The *C*-theory tells us that there is no such thing as time. What we think of as time (drawing upon our experience of it) is in reality something atemporal. Space-time theory tells us that there is no such thing as time; there is only a unity of both space and time. To make us feel 'comfortable' the unity is called 'space-time', a combination of separate terms of which we have a common understanding. If Minkowski had said that there is no such thing as time, but that there is something that, though it appears as time, is something else called, for example, the *X* series, it might have been more accurate. The fact that the terms 'time' and 'space' are both included in the term 'space-time', lead us to think of a combination of space and time, when really we are talking of another entity entirely. In this respect, McTaggart is clearer when he debunks time and calls it something else whereas Minkowski simply says it is only a 'kind of union'[92] and this is not satisfactory as it still leaves us thinking that time is a 'component' of space-time. It is more accurate to think of time as being subsumed into something entirely different as space-time. The space-time cone does nothing but confuse the issue further. It causes us to grapple with a geometrical representation that bears little resemblance to what it is that is actually being proposed.

5. Idealism. McTaggart's idealist system rests upon the perception of spiritual entities he calls selves. Space-time theory is also idealist. It involves necessary perceivers or observers who can never be substantial entities. No person is ever going to observe with any clarity the outer edges of a light-cone. It is also idealist in that it considers 'relatedness' as a unifying function as opposed to the phenomena of that relatedness. Of all the similarities this is the most far-reaching.

6. Reality. The *C* series, by definition, is a *really true-state*, that is not *actually* or *apparently* but ontologically true (*true* and *really true* states are fully described in § 6.6). There is nothing apparent about the *C* series, but logically it is the only reality that lies behind the misleading appearance of time. Space-time is also a *really true-state*, deduced from the Theory of Special Relativity. The Theory of Special Relativity can be subjected to empirical testing, and has at its centre the proof that the appearance of relativistic perception is mistaken.

7. Misperception. The error of misperception is in the *C* series of the perceiver and not in that perceived. Space-time, as a *really true-state*, is only ever misperceived as time by the observer. Though the observer could construe the idea of space-time as interpreting his actual view of time more readily than he could the *C* series.

The two theories come even closer when we consider what Gödel says. Gödel can provide not only a linking 'sub-system' but also a similar ideal world, drawn, not from strictly metaphysical enquiry, but from scientific and mathematical theory. As such, in Gödel we find the culmination of Einsteinian thinking applied to metaphysics. Gödel thinks that the relativity of simultaneity, in the Special Theory of Relativity, gives 'new and surprising insights' into the apparently self-contradictory nature of time. He cites McTaggart's argument against time as the example of this self-contradictory nature.[93] In the Special Theory each observer has his own view of simultaneity and ordering. To one observer events *L* and *M* may be simultaneous and event *N* may be later than events *L* and *M*. To another observer events *M* and *N* may be simultaneous and they may both be later than event *L*. Both observers are correct in what they think. This view can then be shifted back alongside the philosophical demand that time involves change. Neither observer, however, can claim to determine an objective lapse of time. If there is not an objective lapse of time, then change cannot occur. There may be fixed ways of moving from the position of one observer to another, but this system, of co-variance, does not introduce an experience of simultaneity for any observer.

McTaggart denies the reality of time on different grounds. He believes that the temporal terms in the *A* series are contradictory and that change (in the generally accepted sense) does not occur. When 'change' does happen it only happens to events and not to things. As Whitrow points out, '... we can even draw an analogy between the terms of McTaggart's *C* series and the successive backwards-directed light cones with vertices along the world line of an observer in the Minkowski diagram'.[94]

Gödel also points out that the position of matter is not all that straightforward. Observers moving with complete equivalence can do so only in the abstract world of Special Relativity or in some 'empty worlds' of General Relativity. If matter or the curvature of General Relativity is introduced, then the equivalence of different observers is largely destroyed. Inasmuch as although the laws of matter, motion and field are the same for them, the structure of the world may offer a more 'natural' aspect to one observer and a more 'distorted' one to another.[95] This can be answered by the notion of an objective 'world-time' into which fits the different times experienced by all observers in all known universes. Gödel does not accept that such an objective time is valid.[96] He proposes a different cosmological structure. A rotating universe in which the rotational inertia is relative to matter, but is so symmetrical and spatially homogeneous, that any idea of simultaneity or succession cannot be distinguished by any intrinsic properties of the *R*-universe, but only by reference to other *R*-universes. In doing this, Gödel is blending space-time in a way that makes both its 'elements' equally relative. In such a world our epistemic distance to objective matter then becomes as great as it is to objective time.

In doing this, he stands at the same idealist point as McTaggart. Gödel however, sees this relativizing of matter as an opportunity to continue with *A* series thinking. For example, although the logically contradictory nature of time travel is irrelevant, as the practical difficulties (the velocities required to undertake such journeys) make it impossible anyway, Gödel thinks that time travel in an *R*-universe seems possible.[97] Though, unlike McTaggart, Gödel has not denied the existence of time and matter, but blended them into a spatio-temporal whole, we may consider that Gödel and McTaggart share the same intuition. Though I believe that McTaggart's idealism allows him to move forward more freely on his metaphysical quest than Gödel, who cannot free himself from *A*-thinking. Though Gödel considers Parmenides, Kant and McTaggart to be his philosophical precursors, I think he takes too limited a view of what McTaggart has to say. Yourgrau says that on the traditional question of the nature of time, Gödel wakes us 'from our dogmatic slumbers'.[98] Perhaps if we look to McTaggart, and what he offers beyond the denial of the reality of time, we can be shaken out of the waking nightmare of the *A* series.

5.5 The reasonableness of atemporality

Although atemporality goes against our common sense view that everything we experience, we experience in time, I believe we must conclude that it is a reasonable view to hold.

As long as we accept time as real, we must accept the riddles that go with it. The act of communication and our reliance on language continually exacerbate our problems with time. Not only because they occur in an apparently temporal world but also because by in using language to talk about time we get drawn into an unquestioned belief that time (in some way) is real. In talking about time, the fact that it has taken time to talk about it, is always more important than what we have been talking about. Mink thinks that any use of language will always give rise to 'regresses, circles, paradoxes and reduplications'.[99] He says, 'Mr. Collett did not withdraw his remark that Mr. Pickwick was a 'humbug', but his explanation that he meant this remark only in a Pickwickian sense was graciously accepted. Similarly, I [Mink] do not propose that we stop thinking about time, but only that we have wit enough not to argue about what sort of bug a humbug may be'.[100] I think the 'bug' of time will always draw us into the 'humbug' of trying to explain it. But, there is no point in trying to analyse away what is evident. The *A* series is what we use to describe time, it fits our purposes and we use it constantly. To try to explain it away as irrelevant just because it shows itself to be inadequate is absurd. The fact that when it is subjected to closer scrutiny it collapses, should not lead us to find reasons why it should not collapse. This is an apologetic form of philosophy based on a dogmatic belief that time is real no matter how untenable it seems. Instead, we must take a positive step and try to find out what is the true nature of time. Evidently this cannot be found among the remnants of the *A* series nor in the implausible convolutions of the *B*-world.

Because it is not possible for the different stages of the *A* series to be anything other than independent, we suffer irreconcilable problems when we start to bundle them together. This alone makes it perfectly reasonable to seek to replace the time-series. When we add to this all the other confusions we experience on the subject I think we are compelled to look for an alternative.

It is not intuitively easy to deny time. Much of our appreciation of temporality is a socially produced one. Indeed, time may be only a cultural perception.[101] However, it is not essential for survival. It is possible to live in a de-tensed world and still survive. Memory will combine with anticipation to form a (comforting) sense of us 'passing' or 'changing'. But the sense of 'comfort' does nothing to unravel the riddle of time and does not add up to much (if any) certainty anyway. If we look closely at our sensory images of the world and the way we mentally store and catalogue them we can easily feel uncommitted about believing even in our own history. Humean scepticism damns this sort of certainty every time. Added to this is the disquieting suspicion that we are probably the only living creatures on earth that do not live in an almost continuous state of present.[102]

In our everyday lives we can accept the *A* series as a reasonable, if rather rule of thumb, description of the phases of change in time. But although in this way it is inescapable, the fact that it is incoherent and will not yield to any logical test should be enough for us to have misgivings about its reality or about its representing any sort of reality which is temporal. Though there are reasons to uphold the idea of a separate past, present or future there is nothing compelling in any arguments that support them. The idea that the *A* series is made up of incompatible terms, is in many ways, sounder than battling to defend the idea of self-standing parts of that series. McTaggart accepts the pointlessness of such a battle and tries to identify the true nature of what presents itself so confusingly. As far as he is concerned, if the *B* series relies for its reality upon the *A* series and the *A* series is invalid, then the truth can only be found somewhere else than in the *B* series. Although this sits uncomfortably with our everyday experience McTaggart is not alone in his belief (nor is such a view the sole preserve of the metaphysician). As we have seen, Minkowski recognises the same inherent fallacy in the idea of an objective view of time and throws out temporality.

McTaggart thinks that the *B* series is something other than what it appears to be and that what it appears to be is nothing other than an invention of human consciousness. However, all our experience of time is not illusory. When we think we experience a time series, though we are experiencing a series that appears to be temporal and that is an illusion, our experience of something that appears as a temporal series, is not an illusion.[103] In other words, although the *A* and *B* series may not be temporal and are largely erroneous perception they can be 'indicators' of true reality. What really happens in our experience is that we mistakenly apply the apparent terms and relations of the *A* and *B* series to a real atemporal series. McTaggart calls this the *C* series. Although the *C* series is atemporal it is not entirely inaccessible to our temporal experience for even in our misperception of time we are somehow experiencing the idea of the truth of atemporality in the *C* series.[104]

In fact the *A* series, for all its problems, resembles more closely the real *C* series, than it does the *B* series. As McTaggart says, the illusion of time '...consists only in our applying the *A* series to it, [the *C* series] and in the consequent appearance of the *C* series as the *B* series, the relation, whatever it may be, which holds between the terms of the *C* series, appearing as a relation of earlier and later'.[105] It is the *A* series together with the real *C* series that form a primary class that gives us the erroneous experience of time. Both the *A* and *C* series are essential for this erroneous experience though neither of them can be derived from the experience itself.[106] On McTaggart's account, the *B* series however, is not fundamental because it can be derived from the *A* and *C* series as they present themselves to our experience. But if the *C* series is real and its terms are connected by permanent relations (which they must be if it is both real and a series) and it appears to us as an *A* series (which it seems to) then the *C* series must also appear to us as a *B* series. Events in the past will always appear as earlier than events in the future and they will be later than preceding past events.[107] Even so, the permanence of the *B* series does not give it precedence over the transitory *A* series from which it can be derived. The *A* series, more fundamental as it is, is a permanent misperception of the real *C* series.

Even though our experience of time is an erroneous experience of the atemporal *C* series there are some elements of our temporal experience which are not erroneous. If we perceive events as appearing simultaneous, we are in fact having a true perception. These events do occupy the same position in a series though in truth this is the non-temporal *C* series. If we perceive events to be in a time series we will be truly perceiving them in a series but the truth of our perception will lie again, not in a time series, but the non-temporal *C* series. The terms of the *C* series will include everything that appears to us as in time. These apparently temporal events will appear in the same order as events in the *B* series.[108] Although the *C* series may seem identical to the *B* series it is different. The terms of the *B* series are events and the terms of the *C* series are not. The uniting relations of the *B* series are the relations of earlier and later and this is not so with the *C* series. The priority of the *C* series has to be seen in light of McTaggart's metaphysical aims and not directly in relation to time and the *B* series. There is a choice to be made here: the *C* series is a glorified *B* series and as such is an interpretation of a pre-existing real *B* series, or the *B* series is a misinterpretation of the *C* series which itself is real but masked by the *B* series.

Following McTaggart, we are led, not to a denial that what we think of as time exists, nor that it has, as part of it, many features that we normally think of as time. We are led to a notion that this thing that we think of as time, though a series, is not a time series. Though experiencing time is part of our lives we are mistaken if we think that temporal terms will reveal reality. If we need to, we can temper our uncertainty about the fantastic idea of atemporality when we remind ourselves of the fantastically puzzling nature of temporality. Common sense makes us suspicious of McTaggart's conclusion, but it may not be able to restore our faith in temporal reality.

I conclude that time has an inescapable incoherence. But this is itself incoherent for how can we know time to be incoherent? If it is not real then it must at least be appearance, and as such in some way exist. But how can anything incoherent exist? Time cannot be both real and unreal and appear this way as well. This reminds us again that a solution to the problem of the nature of time cannot be found in the terms of time itself and provides us with ample justification for looking at the idea of an atemporal replacement.

5.6 Conclusion

When, in § 5.2, we looked at the position of 'reality of consciousness plus being', we saw that there are strong reasons to believe that McTaggart's denunciation of time should not be dealt with as if it were an end-in-itself. These reasons were strong enough to support withdrawal from the argument over the adequacy or otherwise of the details of McTaggart's argument and its ability to denounce time, and instead move towards a fuller analysis of McTaggart's atemporal replacement. The idea of matching McTaggart's view, first with his fuller concept of the *C* series and then with the modern scientific view of space-time present themselves as natural extensions to what we have found here. Our common sense conclusions in Chapter 4 are re-enforced by this. In § 5.3, when we looked at the position 'reality of becoming plus consciousness', we saw that this position requires an acceptance of temporality. It requires that unless there be something in which change can 'happen' there can be no reality of becoming. Adopting this position therefore throws us back into time's confusing nature. When, in § 5.4, we looked at the contemporary scientific position of 'a four-dimensional universe with no becoming' we came to a surprising conclusion. We saw that McTaggart's idealist view of the *C* series has similarity to Gödel's world structure. On the face of it this seems improbable. However, I identified a number of possible similarities between the *C* series and space-time and concluded that McTaggart's system, because of its complete rejection of temporality, has greater scope. In § 5.5 I concluded that we have sufficient grounds to look for an alternative interpretation of time in an atemporal world.

I think that McTaggart's position is surprisingly close to the views held by science. Indeed, it seems that a 'modern idealism' like this has much to offer the scientific model. It also appeals to some of our common sense values. For these reasons alone it is important to find out what he has to say on the subject of atemporality. But even more than this, McTaggart's argument against time provides us with enough suspicion about the reality of time to encourage us to see what it is he replaces it with. In doing this we go further than most and, in my view, take a decisive and essential step forward. After almost ninety years of inconclusive argument about McTaggart's paradoxical claim about time, we are ready to move to the next stage — timeless reality.

Notes

1 R. Gale, *The Language of Time* (London, Routledge and Kegan Paul, 1968), chs, 10 and 11.

2 Merleau-Ponty, *The Phenomenology of Perception*, p. 412.

3 M. Heller, 'Things Change', *Philosophy and Phenomenological Research*, vol. 52 (1992), pp. 695-704.

4 L.B. Lombard, 'The Doctrine of Temporal Parts and the "No-Change" Objection', *Philosophy and Phenomenological Research*, vol. 54 (1994), pp. 365-72.

5 M. Shallis, 'Time and Cosmology', p. 63.

6 See B.L. Whorf, *Language, Thought and Reality*, ed. J.B. Carroll (Cambridge, Mass, MIT Press, 1956) and E.E. Evans-Pritchard, *Witchcraft, Oracles and Magic among the Azande* (Oxford, Clarendon Press, 1937).

7 See J. Piaget, *The Child's Conception of Time*, trans. by A.J. Pomerans (London, Routledge and Kegan Paul, 1969) and J. Piaget and B. Inhelder, *The Child's Conception of Space*, trans. by F.J. Langdon and J.L. Lunzer (London, Routledge and Kegan Paul, 1956).

8 Geach, *Truth, Love and Immortality*, pp. 100-1.

9 St Augustine, *Confessions*, Ch. 11.

10 J.W. Dunne, *The Serial Universe* (London, Faber and Faber, 1934), p. 9.

11 H. Minkowski, 'Space and Time', lecture delivered at the Eighteenth Assembly of German Natural Scientists and Physicians at Cologne, 21 September 1908, in H.A. Lorentz, A. Einstein, H. Minkowski and H. Weyl, *The Principle of Relativity. A Collection of Original Memoirs on the Special and General Theory of Relativity*, with notes by A. Sommerfeld, trans. by W. Perrett and G.B. Jeffrey (New York, Dover, 1952. First published 1923), pp. 73-91.

12 McTaggart, *The Nature of Existence*, § 347.

13 Ibid., § 349.

14 Ibid., § 348.

15 Ibid., § 351.

16 Aristotle, *De Poetica* (*Poetics*), trans. by I. Bywater, in *The Basic Works of Aristotle*, Ch. 24, 1460^{a}, lines 26-7.

17 H.D. Oakeley, 'Time and the Self in McTaggart's System', *Mind*, vol. 39 (1930), no. 154, pp. 175-93.

18 Ibid., p. 182.

19 Ibid., p. 191.

20 G. Currie, 'McTaggart at the Movies', *Philosophy*, vol. 67 (1992), no. 261, pp. 343-55.

21 M. Dummett, 'A Defence of McTaggart's Proof of the Unreality of Time'.

22 Ibid., p. 498.

23 Ibid., pp. 499-500.

24 Ibid.

25 Ibid., p. 503.

26 King-Farlow, 'The Positive McTaggart on Time'.

27 King-Farlow refers generally to *Some Dogmas of Religion* and all articles reproduced in *Philosophical Studies*.

28 King-Farlow, 'The Positive McTaggart on Time', p. 170.

29 Ibid.

30 Ibid., p. 171.

31 Gale, *The Philosophy of Time*.

32 Gale, *The Language of Time*, pp. 11, 94 and 95.

33 Ibid., p. 11.

34 See Gale, 'The Static Versus the Dynamic Temporal'. For an example of a 'standard' *B*-theory answer to McTaggart see pp. 70-7, and for *A*-theory see pp. 71-83.

35 D.C. Williams, 'The Myth of Passage', *Journal of Philosophy*, vol. 48 (1951), no. 15, pp. 457-72.

36 King-Farlow, 'The Positive McTaggart on Time', footnote, p. 174.
37 Grünbaum, 'The Status of Temporal Becoming'.
38 King-Farlow, 'The Positive McTaggart on Time', p. 177.
39 King-Farlow discussed his ideas with W.R. Shea and says Shea helped him broaden his initial perspective on McTaggart. (Ibid).
40 Yourgrau, *The Disappearance of Time.*
41 Wittgenstein, *Tractatus Logico-Philosophicus*, § 1.1.
42 Ibid., § 7.
43 Bergson, *Time and Free-will.*
44 E. Husserl, *The Phenomenology of Internal Time-Consciousness*, trans. by J.S. Churchill (Bloomington, Indiana University Press, 1964).
45 Merleau-Ponty, *The Phenomenology of Perception*, ch. 2, 'Temporality'.
46 M. Heidegger, *Being and Time*, trans. by J. Macquarrie and E. Robinson (New York, Harper and Row, 1962).
47 Merleau-Ponty, *The Phenomenology of Perception*, p. 412.
48 Ibid.
49 See Mellor, *Real Time.*
50 A.J. Ayer, *Philosophical Essays* (London, Macmillan, 1954), ch. 7, 'Statements about the Past'.
51 Ibid., p. 102.
52 G.N. Schlesinger, 'E Pur Si Muove', *The Philosophical Quarterly*, vol. 41 (1991), no. 165, pp. 427-41.
53 J. Biglow, 'Worlds Enough for Time', *Noûs*, vol. 25 (1991), pp. 1-19.
54 D.J. Buller and T.R. Foster, 'The New Paradox of Temporal Transience', *The Philosophical Quarterly*, vol. 42 (1992), no. 168, pp. 357-66.
55 See Kant, *Critique of Pure Reason.*
56 See Aristotle, *Metaphysica (Metaphysics)*, trans. by W.D. Ross, in *The Basic Works of Aristotle*, pp. 681-926 and *De Anima.*
57 S. Rosen, 'Thought and Touch. A Note on Aristotle's *De Anima*', in Yourgrau, *Demonstratives*, pp. 185-94.
58 See W.V. Quine, 'Three Grades of Modal Involvement', in *Ways of Paradox* (New York, Random House, 1966), pp. 156-74.
59 Lucas, 'The Open Future', p. 131.
60 L. Sklar, 'Time, Reality, and Relativity', in Yourgrau, *Demonstratives*, pp. 247-60.
61 C. Rietdijk, 'A Rigorous Proof of Determinism derived from the Special Theory of Relativity', *Philosophy of Science*, vol. 33 (1966), pp. 341-4.

62 H. Putnam, 'Time and Physical Geometry', *Journal of Philosophy*, vol. 64 (1967), pp. 240-7.
63 R. Teichmann, 'Time and Change', *The Philosophical Quarterly*, vol. 43 (1993), no. 171, pp. 158-77.
64 Shoemaker, 'Time Without Change'.
65 Ibid., 177.
66 Sklar, 'Time, Reality, and Relativity', p. 255.
67 Putnam, 'Time and Physical Geometry', p. 246.
68 Sklar, 'Time, Reality, and Relativity', p. 258.
69 See: C. Glymour, 'Topology, Cosmology and Convention', *Synthèse*, vol. 24 (1972), nos. 1/2, pp. 195-218; C. Glymour, 'Indistinguishable Space-Times and the Fundamental Group', in J. Earman, C. Glymour and J. Stachel, eds., *Foundations of Space-Time Theories, Minnesota Studies in the Philosophy of Science*, vol. 8 (Minneapolis, University of Minnesota Press, 1977), pp. 50-60; D. Malament, 'Observationally Indistinguishable Space-times: Comments on Glymour's Paper', in Earman, Glymour and Stachel, *Foundations of Space-Time Theories*, pp. 61-80.
70 Welby, 'Mr McTaggart on the "Unreality of Time"'.
71 Ibid., p. 326.
72 Ibid.
73 Ibid., p. 327.
74 Ibid., p. 328.
75 N. Markosian, 'How Fast Does Time Pass?', *Philosophy and Phenomenological Research*, vol. 53 (1993), pp. 829-44.
76 Shea, 'McTaggart and the Neo-positivist Entropists'.
77 In particular, Reichenbach, *The Direction of Time*.
78 In particular, A. Grünbaum, *Philosophical Problems of Space and Time*.
79 McTaggart, *The Nature of Existence*, §§ 595-604.
80 Ibid.
81 Shea, 'McTaggart and the Neo-positivist Entropists', p. 347.
82 Ibid., p. 348.
83 Reichenbach, *The Direction of Time*, p. 137.
84 Ibid., pp. 135-43.
85 Ibid., p. 133.
86 Grünbaum, *Philosophical Problems of Space and Time*, pp. 257-64.
87 Shea, 'McTaggart and the Neo-positive Entropists', p. 350.
88 Eddington, *The Nature of the Physical World*, p. 35.

89 J.V. Luce, *An Introduction to Greek Philosophy* (London, Thames and Hudson, 1992), p. 54.

90 McTaggart pays scant regard to science and because of his commitment to idealism has little time for the empirical method. He was, however, aware of Einstein's work and supported the idea of multiple time series. In part of his description of the *C* series he provides us with ample reason to proceed with lining up the *C* series and space-time. (McTaggart, *The Nature of Existence*, § 527).

91 L. Sklar, *Space, Time and Spacetime* (Berkeley, Calif., University of California Press, 1977), p. 2.

92 Minkowski, 'Space and Time', p. 75.

93 Gödel, 'A Remark about the Relationship between Relativity Theory and Idealistic Philosophy', p. 261.

94 G.J. Whitrow, *The Natural Philosophy of Time* (Oxford, Clarendon Press, 1980), p. 348.

95 Gödel, 'A Remark about the Relationship between Relativity Theory and Idealistic Philosophy', p. 262.

96 Ibid., p. 263.

97 Ibid., p. 264.

98 Yourgrau, *The Disappearance of Time*, p. 167.

99 Mink, 'Time, McTaggart and Pickwickian Language', p. 263.

100 Ibid.

101 M. Shallis, 'Time and Cosmology', in R. Flood and M. Lockwood, eds., *The Nature of Time* (Oxford, Basil Blackwell, 1988), pp. 63-79, p. 63.

102 See W. Kochler, *The Mentality of Apes*, trans. from the 2nd revised edn., by E. Winter (Harmondsworth, Penguin, 1957) and S. Walker, *Animal Thought* (London, Routledge and Kegan Paul, 1983).

103 McTaggart, *The Nature of Existence*, § 347.

104 Ibid., § 349.

105 Ibid., § 347.

106 Ibid., § 348.

107 Ibid.

108 Ibid., § 351.

6 McTaggart's Timeless World

6.1 Introduction

In this Chapter I describe the atemporal world that McTaggart calls the *C* series. I show that this can lead us to a plausible (if not convincing) explanation of what time may really be.

In § 6.2 I describe the replacement for time, the atemporal *C* series. In § 6.3 I describe how this is coherent with McTaggart's description of reality. In § 6.4 I describe how the *C* series is not just an alternate description for what we think of as time, but can illuminate our puzzled view of time. I do this by lining up the 'inclusion series' with the asymmetrical nature of time (§ 6.4.1) and its directionality with the apparent direction of the *B* series (§ 6.4.2). In § 6.5 I describe some fundamental relations between the temporal world as we experience it, and the atemporal world of the *C* series. In § 6.6 I look more closely at the distinction between apparent and actual time. 'Apparent time' is the time in our experience that does not seem to resemble what time may really be (many of our views on time are like this). 'Actual time' is the measurable time of our everyday world. Actual time and apparent time I call '*true*' time and the atemporal order that in McTaggart's idealist world is reality I call '*really true*' 'time'. In this way we can keep the confusing notion of our experience of actual time as *true* time, separate from a plausible explanation of what it may really be in *really true* 'time'. In § 6.7 I conclude that it is reasonable to look beyond McTaggart's claim that time is unreal. However, his atemporal world is not easy to accept in total. Indeed, this is at the root of the problem. Because he proposes a complete metaphysics we get easily bound up with his ontological quest and may find ourselves rejecting atemporality for the wrong reasons.

6.2 The atemporal C series

In this section I describe McTaggart's replacement for time, the atemporal *C* series. Fully to appreciate the *C* series, it is necessary to consider additional features of McTaggart's fuller purpose (see § 2.4).

As we have already seen the *B* series depends upon the *A* series and the *A* series demands past, present and future. If events do not appear in the *A* series they cannot be in the temporal *B* series. Unless they change, they cannot be in time and they can only change if they involve the *A* series. But the misperception that we have is of something appearing to us in the *A* series. This misperception, in itself, gives us no indication as to where the past or future lie relative to the present, or upon which side of the present any events are that may be past or may be future. But there is something about the experience that implies a sort of 'passing along' the *A* series. (McTaggart thinks that the *A* series 'slides against' the *B* series in some way). This asymmetry is not satisfactorily answered in the terms of time. However, the fact that the misperception produces the idea of an *A* series means that we are being aware of some sort of series. Our awareness of such a series must mean that there is something on one side or another of any given term. It must also mean that the given term is neither closer to it nor further away from it than any other term on the same side of that given term. This series is the *C* series.

We can never directly perceive the *C* series. We can only infer its existence from the reality which we misperceive as the *A* and *B* series. Although time is unreal the *A* and *B* series still exist as misperceptions. The *A* and *B* series depend upon the *C* series, and because the *C* series is real, they exist, whether or not we mistakenly perceive them. However, if we did not misperceive the *A* and *B* series they would not in any other way appear to us to exist.[1] That which we misperceive as the *A* and *B* series causes an incorrect inference about the nature of the *C* series and creates, in our minds, a false idea of something that does not exist at all. The *C* series is therefore the creator of an unavoidable misapprehension of time in the *A* and *B* series.

Whatever *G* perceives in *H* it is not in the *A* or *B* series because neither of these is real. Because the perceptions are in *G* (the perceiver) and not *H*, and because these are misperceptions of time, it follows that there must be as many time-series as there are selves perceiving things in time.[2] This leads to many time series being apparently similar. Two percipients may assume that they are perceiving the same moment in one time and this can lead to our wrong (though convenient) conclusion that there is one common, objective time.[3] In the *C*-world the idea of a common time is based upon common misperception.

If *G* perceives *H* in three successive states, black, red and white, then this forms a *C* series in *H*. The *C* series of *H* cannot account for the misperceptions that *G* has of *H* of *H* being in time or being a material object because these are false and the *C* series of *H* is true[4] (because it is a passive series of 'being perceived' and therefore precludes misperception). The root of misperception is therefore in the 'actively perceiving' *C* series of *G*. Error is therefore to be found in the perceiver and not that which it perceives. Consequently, time is a false product of the act of our perceiving.

On the face of it the *C* series simply seems the system of *determining correspondence*, but this is not so. The *C* series is one dimensional and *determining correspondence* is two-dimensional: one dimension being the terms of the grades of perception and the other the terms of the parts of those grades. In *determining correspondence*, one part is determined by another part and that determining part includes as part of itself the determination of the other part. In fact, it would be possible to arrange the *determining correspondence* series in one dimension but there still would be no definite place for parts in it, whereas in any *C* series everything has a definite place. The *C* series must be one dimensional because it is a simple series of perception. In the *determining correspondence* system a self can only have one perception of any one perception whereas in the *C* series (because of the system of *determining correspondence*) a self can have many perceptions of the same perception at different times.[5]

However, our experience is not that we have perceptions in such a detailed way. We do not, after all, experience perceptions that differentiate parts within parts to infinity. Nor can we be fully aware of the extended range of the *C* series which by *determining correspondence* is connected to the *C* series of other selves. Added to which, all our perceptions are confused by appearing as if they are in time. These perceptions — that are not clear and distinct but form the differentiation of parts to infinity — are *fragmentary parts* (McTaggart represents them by the symbols c_1, c_2 and so on). Whereas *determining correspondence* parts must be distinguishable from one another (by differentiation of parts to infinity) this is not so for *fragmentary parts* which may be highly confused. Because all *determining correspondence* parts can be fully described *determining correspondence* cannot deal with time. Time is therefore described by the system of *fragmentary parts*.

Every *determining correspondence* part of a self has a sufficient description, for example, *G!H*. However this is not sufficient to describe the terms of the *C* series which includes the confused temporal perceptions of the perceiver. To completely describe a *C* series we are therefore looking for a description $c_1G!H$, or $c_2G!H$ and so on.

G!H is a correct perception but all parts of *G!H* in the *C* series are misperceptions. *G* misperceives, as a *B* series, terms of *H* in the *C* series c_1H, c_2H and so on. All selves have correspondent and different *C* series therefore when *G* perceives as present *H*, *H* is not sharing the same *C* series 'present' as *G*.[6] This is, however, confused by the fact that *G*'s apparent perception of *H* in the present involves both of them being at the same stage in the *C* series though this stage is only a stage involving the *fragmentary parts* of time.[7] A self-perception *G!G* is also misperceived as in time.

Although *G!H* is a correct perception of *H*, the *fragmentary parts* of *G!H* in the *C* series are misperceptions of *H*. This means that any perception we have includes that which is both real and unreal. We perceive others as material when they are not and we perceive them as in time when they are not. The misperceptions that we have are nevertheless part of our true perception. For example, the perception *H!K* is the perception of *K* as a spiritual self[8] and *G!H!K* is a perception of *H!K* having a perception of *K*. However, there are parts of a self (the confused *fragmentary parts*) that sometimes perceive *K* as material. If this is so then *H!K* would be perceiving *K* as both material and spiritual which is, of course, impossible.[9] Neither can *H!K* be perceived occupying a different position in the time series than its *fragmentary parts*. Parts of a whole cannot occupy a different place in a time series than the whole itself. Accommodating error into the system seems to expose it to internal contradiction.

The situation is this. Although time is derived from a plurality of *fragmentary parts*, those *fragmentary parts* together form a whole, which by itself, can only be one term in the dimension of time. But one term alone cannot appear in time because of the *B*-feature demands of earlier and later.[10] This implicit contradiction is because of the implicitly contradictory nature of time itself. *Fragmentary parts* are part of the system of misperception and must be discounted when we talk of the nature of reality. Because this error in our perception cannot be found in the system of *determining correspondence*, the error that we experience when we perceive *fragmentary parts* does not lead to there being error in the system.

This is a little confusing and involves some manipulation of the interdependence of wholes and their parts. It is at the crux of McTaggart's argument to replace time and, on the face of it, it looks like he is trying to get the best of both worlds. He is faced with having to accommodate our obvious experience of time (which is erroneous) into a timeless framework that does not include error. Even though the *fragmentary parts* of time are consigned to a less important realm, it is difficult to see how *determining correspondence* still does not include them. We may be left thinking that although *fragmentary parts* are necessary in the system of *determining correspondence* they cannot be conveniently bundled together when they get in the way. It may be that we could assume that if clear and distinct perception are necessary for perceptions, such as *G!H*, then *fragmentary parts* have no role anyway. But we are then left with the obvious confusion of perceptions and the way this excludes our clear perception of parts of a self to infinity. There is something of a 'leap of faith' here. It is as if the universe is discovered to contain firstly nothing which is green and secondly no misperception in which case we would have to wonder how our experience of greenness can possibly be part of the universe.

The problem of error and erroneous perception stalk the *C* series wherever it goes. McTaggart says that what is perceived exists and exists as it is perceived subject to the condition that, '*A* exists and is *X* at a point or points in the *C* series that appear to be present at the point in the *C* series at which the perception exists'.[11] In other words, 'When I perceive it'. However, we are left to wonder if this position allows for erroneous perception. McTaggart's argument is that no part of *determining correspondence* can perceive anything as in time. Although perceptions in *determining correspondence* may not represent everything that is contained in the object of perception, it cannot perceive any characteristics that are not there to be had. No perception in the system can be erroneous. In the same way, although our perception of the *C* series may be erroneous, all perceptions in the *determining correspondence* system must be without error because it is a system that precludes error. We may find this improbable and think, like Geach, that, 'we cannot without contradiction say that we *only appear* to fall into error, endure pain, or enjoy pleasure, and do not really do so'.[12] However, this is not exactly what McTaggart is saying. He accepts that we experience the pain but does not accept that it forms part of a reality that is made up only of 'unpainful' experience. The fact that 'unpainful' experience includes pain but that it is not real pain because it is not part of reality is the problem here. The apparent experience of pain (or time), although an experience, can certainly be consigned to the confusion of misperception even though still we may feel their effects. This sort of pain is more psychosomatic. Even though the experience seems real enough it is not really pain only our perception of mental confusions that cause us to think of pain. Translated to time, our experience of time is imaginary and though we are having a real imaginary experience this does not bring any reality to time. *Determining correspondence* can include clear and distinct perceptions of imaginary things like time, without having to deal with any problems associated with the object of that imagination.

The link between misperceived time and real timelessness is real enough so now we must look at how the *C* series resembles the *B* series. The relations of the *C* series must reflect the relations of the terms of the *B* series. In fact, the relations of the *C* series *are* the relations of the *B* series as it appears to us. Therefore the *C* series must in some way resemble the transitive, asymmetrical relation of earlier and later.[13] In addition, this must apply to every *C* series, not just in the terms of the *C* series of one self.

According to McTaggart a whole is made up of a set of parts that are neither more nor less than the set of parts that makes it up.[14] None of these parts can be added in twice nor can they overlap. On this basis, all the content of *G!H* falls within the terms of the *C* series of *G!H*. But the *C* series does not have to be a set of parts of *G!H*. For example, (McTaggart's pre-metrication example) a foot ruler that can be divided into twelve one inch parts, all of which will contain its content and be a set of parts of it, is made up of a set of parts of one inch, two inches, three inches and so on, up to eleven inches. The twelfth part would be not only twelve inches, and the next in the series, but would also be one foot. The parts would be part of a series that contained all, and nothing other than, the whole content of the foot. It would not be a set of parts of the foot because together they add up to more than a foot and, except the one of them that is twelve inches, all the others could be subtracted without affecting the whole[15] (inasmuch as there would be at least some of the whole remaining).

But the *C* series cannot easily be represented by these means because the terms of the *C* series of *G!H* *are* a set of its parts. When the different terms $c_1G!H$, $c_2G!H$ and so on are added, they will form *G!H*. But as $c_1G!H$, $c_2G!H$ and so on are all misperceptions, it would be incorrect to think that simply adding them together somehow makes them into a correct perception.

These confused terms appear in the *B* series as the relations 'earlier than' and 'later than'. According to McTaggart, these relations are more rightly viewed as the *C*-terms 'included in' and 'inclusive of'.[16] As *B*-terms can be arranged in a definite order from earlier to later, so *C*-terms can be arranged in a definite order of 'successive inclusion'. Because *C*-terms can be substituted for *B*-terms in this way it is superficially reasonable to assume that the terms 'included in' and 'inclusive of' appear as the relations 'earlier than' and 'later than'.

Although the relations generally present themselves as pairs (earlier than-later than) they are in fact separate relations, 'included in' appearing as 'earlier than' and 'inclusive of' appearing as 'later than'.[17] McTaggart assumes that the *C* series does not possess any other relations that could appear as *B*-relations.[18] Terms in the *C* series also differ in magnitude, more inclusive terms having greater magnitude than less inclusive terms. This sort of 'intensive' difference (as McTaggart calls it) in magnitude is not a difference in perception. It is more like the difference between one state of pleasure and another not being itself a third state of pleasure.[19]

Intensive difference also involves extensive difference because 'more inclusive' implies some sort of increment. These increments will themselves form a series that has extensive magnitude. This series, which would then provide a system for quantifying the difference between intensive differences in magnitude, is not included in, nor does it include, any other. McTaggart calls this extensive series of increments the *D* series. The *D* series is not part of the *C* series but it is by the *D* series that we recognise our misperception of the *C* series as the *B* series.

It seems puzzling how 'more inclusive' implies an increment of some sort. However, the relation of the *D* series to the *C* series does not present the same sort of problem that the relation of *fragmentary parts* to *determining correspondence* presents and is easier to accept. I think the answer is in the way the *C* series builds on mutually held perceptions. Although most perceptions will be erroneously in time and therefore *D*-perceptions of the *B*-world, some perceptions will be true. For example, a self's perception of love for another self will add to the inclusiveness of the self and because each addition adds more it must be incremental. The value of love for the other self will build on the value it had before the latest perception. Most of the contents of the world are misrepresented in time or are not part of the self's *C* series and so would not have this building effect. For example, that Picasso's blue period paintings are more blue than Van Gogh's does not add anything but as the comparison is both temporal and not part of a self's *C* series it is ruled out as an increment of inclusiveness.

If the *C* series has a first term then the *D* series will also have a first term. The first term of the *D* series will be a state of misperception (that is an increment in the amount of perception). It will also be a member of the set of parts that are parts of correct perception. This is not contradictory (as it at first seems) because every state of misperception in the *C* series is also a member of the set of parts of correct perception.[20] The *D* series cannot be in respect of increased clearness of perception because this clarity 'oscillates' because of the confusion of *fragmentary parts* (that is, sometimes we perceive more objects and sometimes we perceive less, consequently there is a change in the extent of our experience[21]) and the *D* series is a uniform asymmetrical change. It must therefore be an additional perceptual framework. The nature of the *D* series must be in the percipient *G* because it is to do with the increased perception of *H* as a whole. The *D* series therefore involves no additional perception though, in some way, according to McTaggart, everything that we perceive in it, we must perceive more.[22] I take it that he means that although increments of the *D* series do nothing to add to the perception *G* has of *H*, the increments of the *D* series are themselves additions to the perceptual framework as a whole. If this is so then it seems reasonable and does not appear to violate the system's believability.

The situation is this. The terms of the *D* series and the *C* series are not known to us by perception although we do perceive the *C* series as the *B* series.[23] Because of our perception of the *B* series this means that the three series all have psychological importance. Together the three series constitute a third dimension adding to the two dimensions of the *determining correspondence* system. The *C* series, because it is an inclusive series, includes some states of misperceptions in a misperception series (for example, some terms of the *C* series appear as a *B* series to some percipients and therefore constitute a misperception series) and this is the *D* series.

'Reality', the perception of *H* as a whole, can only occur in the *C* series, it can never form part of a misperception series. Everything in all our pre-final (temporal) experience is in a misperception series because all our experiences are in time.

The terms of the *D* series are 'amounts of perception' and will always form part of the states of perception of the *C* series. There is no possibility of there being another way in which *G!H* perceives *H* because all terms in the inclusion series are *fragmentary parts* of *G!H* or are misperceptions.[24] Geach claims that *G!H* does not include *G*'s perception of *G!H*.[25] I see no reason why not. It is only parts of *G!H* that are a misperception and the whole is a correct perception. When Geach says that *G!H* does not contain 'the whole content absolutely of the person *G*', I think he is using 'absolutely' in some sense that is extra to the 'whole' content of *G*. In extending the sense of *G* in this way, he makes room for more perceptual possibilities that, on McTaggart's account, are not available. If *G* is the content of the whole (when all perceptions are had in the last term) then there is no possibility of some 'absolute whole' which is more than this.

For McTaggart the total of all perceptions must amount to the whole. He rejects the idea that in the *C* series there could be various terms besides *G!H* as a whole (as the last term). After all, *G!H* is the whole content *G* as it contains all its perceptions, both of itself and of *H* (and this includes its own self-perception).[26] Geach thinks that any distinct states of perception separate from *determining correspondence* involves an additional misrepresentation of the self's perception in the *C* series. He is not convinced that c_xG, c_yG and so on, can be distinct states of perception unless they misrepresent themselves as having content separate to *G*. But we have already seen that the misperception series (the *D* series) is an inherent part of the *C* series. The *C* series is not just the system of *determining correspondence* though it includes it. Geach's idea that there is something wrong with the *C* series if it cannot include stages distinct from perception of *determining correspondence* ignores this. The *C* series still can be infinitely divisible by *determining correspondence* and still accommodate the increments of the *D* series. Because of this it can include stages of misperception distinct from perceptions in *determining correspondence*. Geach also thinks that a self who lacks self-perception cannot perceive misrepresentations of any *C*-stages themselves.[27] But there is no point in imagining a self who lacks self-perception; no such self can exist in McTaggart's system because the self is the quality of perceiving itself. A self who perceives itself has no *fragmentary parts* (as perceptions)[28] because *fragmentary parts* can only be misperceptions of relations one to another. It would have an infinite series of perceptions in the system of *determining correspondence* but not any *fragmentary parts* and therefore no misperceptions (all of such a self's perceptions being correct perceptions). These misperceptions, cannot be separate to the inclusion series unless they appeared to themselves to have a content not shared by other perceptions and this is not so. There can only be one difference between the inclusion series and the misperception series. That is when the last term in the inclusion series (the term of 'fully inclusive') occurs because this last term is not a term in the misperception series.

McTaggart says the nature of the misperception series precludes the possibility of *G* perceiving any parts of the series as not in time. This is because all misperceptions of 'included in' or 'inclusive of' occur in the *B* series.[29] He thinks it unreasonable that there are additional misperceptions of 'included in' or 'inclusive of' that hold relations that are at the same point (in the series) different from 'earlier than' or 'later than'. *H* as a whole will always be perceived of as in time, either as the latest term in the future or the earliest term in the past although *H* as a whole will never be perceived as present. The only term in the *C* series of *G* that is at the same stage as *H* as a whole is the term *G* as a whole. Only in this respect can there be judged to be such a thing as a common time series.[30] Though of course, we mistakenly conclude that there is a common time series in the *B* series.

McTaggart says it is possible that a self should perceive either its own inclusion series or the inclusion series of another self as not in time.

However, what sort of terms would apply, that would replace the term of the *B* series, cannot be known.[31] Such a set of terms would exclude one another and could not be time itself nor dependent upon it. The inclusion series has terms that are transitive and asymmetrical and, at the same point in the common inclusion series, transitive and symmetrical. The time system has corresponding dimensions, the transitive and asymmetrical relations of the *B* series and the transitive and symmetrical relation of simultaneity. For the *C* series to make sense, it must involve these two pairs of relations.

According to McTaggart, the alternative cannot be found in spatial relations. Spatial connectedness cannot allow for relations that are directly between the terms in the way that earlier than-later than are. Though he believes the two dimensions needed may be found in the points in a finite space to something outside them (for example, the points within a square to one of its sides), they cannot be found without that external relation (in this case, of the side).[32] Neither can an alternative be found in changes in objects, in clearness or accuracy of apparent knowledge or in causation. None of these characteristics show any 'uniform change in time' because they all oscillate. As oscillation occurs in time, they would have to oscillate in the *C* series. The *C* series is an inclusion series and if it is perceived in any other way than as time any oscillation would have to occur in that other series. But, according to McTaggart, no other series will support *B*-relations. For him, no other system except time, will support the misperception of inclusion in the *C* series.[33]

The final stage of the *C* series will always appear *sub specie temporis* as future (though not of course as transitory because there can be nothing that follows it). Because it is seen from within a misperceived world of temporality, it is always seen as an unending duration. But if no self is in time then it cannot endure endlessly. If however, being immortal combines 'not ceasing to exist' and 'appearing to exist in an endless future' then all selves are immortal. The *C* series provides the framework within which both conditions can be fulfilled.

I conclude that atemporality allows us to interpret our existence in a way that the puzzling nature of time constantly confounds. It provides a 'replacement' for time and a positive framework for atemporal existence. In doing this, it largely sets aside the temporal puzzle as we experience it.

6.3 Coherence of the atemporal series

I now go on to describe how the atemporal *C* series is a serious attempt to replace the *B* series and does not form part of the argument for the denial of time. The *C* series is coherent with the relevant elements of McTaggart's fuller philosophy. In this sense its basis lies here more than on the denial of the reality of time itself.

McTaggart lays down twelve conditions that need to be met in order for the *C* series to be coherent.[34] I follow each of his conditions with a brief review.

1. The *C* series must be infinitely divisible by *determining correspondence* as this is the only way that substance can fully be described.

We have seen that the system of *determining correspondence* raises questions which are not easy to resolve. However, the system of *fragmentary parts*, even though this too is problematic, helps answer the concern about confused perception and the place of misperception in a series which precludes error. Although the system of *fragmentary parts* only works if the nature of the self, as a perceiving thing in a non-material and atemporal world, is accepted.

2. Following on from '1', it is obvious that the *C* series must adapt to both time and erroneous perceptions. It must also reconcile with our *prima facie* certainty that when we have perceptions, they are real.

The apparently true perception that occurs, 'at the time that it is made' is also the true perception, 'at the point in the *C* series at which it occurs'. All perceptions in the *C* series are misperceptions except the final stage in the *C* series, which is *G!H* as a whole. We know that the amount of error or misperception in the *C* series varies with the amount of perception. But there are also oscillating differences in degrees of perception as well as error at all points in the *C* series. Although every perception will be, in part (to some degree or another) misperception, it may be the case that still we are perceiving something of the true characteristics of the quality and relations. If this is so then even erroneous perceptions are perceptions of something real.

Because every perception, except *G!H* as a whole, is a misperception, any perception in time must be a misperception. But that misperception still has some relationship or link with the object of perception in the *C* series. There is therefore some 'truth' in the misperception of something in the *B* series because, in some way, it resembles or is identified with a true perception in the *C* series. If we can accept this then it does not matter that all perceptions in the *C* series, except *G!H* as a whole, are misperceptions because all misperceptions are 'subordinate to' (or subsumed by) the true perceptions in the *C* series. In this way, we are not, as Geach points out, 'just under the illusion of being under illusions'.[35]

3. The *C* series must adapt to the sort of erroneous perceptions that exist, for example, matter, sensa, judgements, assumptions and so on.

McTaggart has told us that all perceptions that seem to be matter, sensa, judgements, assumptions and so on are really misperceived objects of our perceptions. By adopting McTaggart's definition of substance we are assuming that there are, in truth, no such things as matter, sensa, judgements, assumptions and so on (only our erroneous experience of their existence). All that exists is spirit that has no content except perceptions. Indeed, according to McTaggart, even the relation that perceptions seem to have to emotional and volitional qualities are not certain.[36] If there are no such things as these then we must be having erroneous perceptions when we think we perceive them. Even judgements we make about the nature of error in our perceptions must be in doubt. As all our perceptions are in the *C* series the *C* series must fit such erroneous perceptions. But it is difficult to see how this can be so. The whole *C* series apparatus only works in an ideal world consisting in spirits, their perceptions and nothing else. Error, whether apparent or real, or judged correctly or incorrectly, is part of our perceptual experience and if all our perceptions are in the *C* series (and this is without error) we are rather cast adrift. This is a clear weakness in McTaggart's system and can only be answered by allowing (as McTaggart does of evil) that the *C* series is without error in a 'relative' sense, inasmuch as when it reaches the final stage the amount of error that it has perceived will be so small compared to the amount of clear perception that it will be rendered insignificant (if not harmless).

4. The *C* series must be transitive and asymmetrical in the same way as the apparent *B* series, for if the *C* series appears as the *B* series it must have its apparent relations.

We have seen that the relations of the *C* series are either 'included in' or 'inclusive of' and either of these relations is both transitive and asymmetrical.[37] This explanation acceptably translates temporal asymmetry to an atemporal serial asymmetry.

5. The *C* series must have as many terms as are necessary to distinguish all the terms of the *B* series and any more that we do not perceive. This may mean that it has infinite or finite terms.

If this were not so then there would be no explanation for the *B* series as a misperception of the *C* series. But there is nothing in the theory that prevents the number of terms in the series being as great or small as is required. The series must all fall within the whole *G!H* but there is no limit to the number of increments of the *D* series that fall between the terms of the *C* series. The *C*-system does not prescribe any minimum increment in the *D* series. We do not really know whether the terms of the *B* series are next to each other anyway. It is only our convenient 'packaging' of events in time that leads us to create a mental picture of causal contiguity. Because the *C* series does not suffer from this sort of 'packaging', the terms of the increments of the *D* series may or may not be next to each other. If there are minimum increments in the *D* series, then they would be next terms in the *C* series. If there were no such minimum increment, there would, of course, be no next terms in the *C* series.[38]

6. The *C* series must allow for the fact that in our experience we only perceive part of that which makes up absolute reality, which by *determining correspondence* is an infinite differentiation of parts.

In our experience we may perceive some parts but we do not perceive an infinite number. Nor do we perceive all the past and future parts of the parts of *determining correspondence* that we do perceive.

We cannot say that what we do not perceive now, we have perceived at a past time, or will perceive at a future time. Whatever exists at any point in the *C* series exists as a part that, by *determining correspondence*, can be divided infinitely. What I perceive is, in fact, infinitely divisible, although it does not appear so. The misperception of *H* (that is the confused perception of *H*) that includes all its *determining correspondence* parts (McTaggart represents this by '*H+*') does not have to be an infinite number of perceptions itself. Any number of these infinitely divided parts may only form one perception. Our perceiving of only the one perception does not preclude the existence of all the parts of *determining correspondence* in *H+*. This could not apply to the last term of the *C* series *G!(H+)* which of course is not a fragmentary part but is *G!(H+)* itself. *Fragmentary parts* are themselves misperceptions and so the misperception of them raises no problem. *Fragmentary parts* may form parts of aggregated perceptions such as *H!K*, *H!M*, *H!Q* and so on, but this makes our misperceptions of them no less tenable.[39]

7. We must be able to account for the persistence and recurrence of our experience of a time series so that judgements and assumptions we make about occurrences in the *B* series are accounted for.

The stages in the *C* series are dissimilar to one another. They can, however, also be similar because at every stage all parts of each self will be infinitely divisible. Because our perceptions of stages in the *C* series are only partially erroneous our successive perceptions will tend to resemble one another. Also, because the nature of the error must change so that different stages in the *C* series are dissimilar, there need not be any complete change at any particular stage. In this way our erroneous judgement of a time series will persist.[40]

8. The *C* series must account for the appearance that there are more objects of experience at one time than at another. Any increase in apparently perceived objects is followed by a decrease. Because of this, there is an oscillation in the apparent quantity of experienced objects. It must also accommodate changes in the varying clarity of our experiences and the fact that in dreamless sleep we appear to experience nothing.

Because we have perceptions from 'included' to 'inclusive' in the *C* series there is an implicit increase in the amount of perception. If, on the other hand, our perception went from 'inclusive' to 'included' there would be a commensurate decrease in perception. But such increases and decreases would not necessarily mean associated increases and decreases in clarity of perception, particularly when viewed in an intensive (as opposed to an extensive) way.

There is a quantitative change as we proceed from 'included' to 'inclusive' but this does not necessarily mean that it is only quantitative. Qualities may be added and if so, although there is a quantitative increase in perception, the additional qualities may need further additional qualities to allow the perception to be increasingly clear. This may mean that increased perception leads to increased confusion that will only become more clear as more perception is had.[41] On this basis, oscillation of clearness could occur at any stage and would therefore appear as an oscillation in time. In addition to the external effect of time, some oscillation of perception also occurs in the percipient. This can lead to two percipients perceiving the same event differently. Oscillations will occur as certain characteristics appear as differences in events and such differences will affect our clearness of perception.[42] McTaggart thinks that in the case of dreamless sleep we are forced to conclude that the *C* series passes, without an intermediate state, between the last and the subsequent first moment of consciousness.[43] This does not mean that we do not exist when we are not conscious but it does mean that there is no increment in the progress of the *C* series — our conscious universe at this time 'stands still'.

There is also the problem of oscillations in the extent of perception. On the face of it, our perceptions at one moment may be great in number and in the next very few. If this were so then it would seem to suggest a great reduction in the extent of perception. This is puzzling because, unlike dreamless sleep, if we are conscious we should be conscious of perceiving everything that we have previously perceived (because the *C* series is a relentlessly building series which does not allow for reduction). All the differentiating groups and all its parts are represented at any point in a percipient's *C* series, so the extent of perception ought to be the same at all those points. According to McTaggart, this is really the same as the problem of our perception of only parts of what, in fact, is an infinite differentiation of parts. What seems an oscillation in the extent of a field of experience, is only an oscillation in the proportion of that field that was above a certain level of clearness.[44] But this is not a complete solution. Sometimes we know that we can perceive a 'field of objects' without perceiving separate objects — we can see the wood without seeing the trees. McTaggart thinks that our 'perception of a vague background to our more definite perceptions', provides us with an idea of what perception is like without perceiving definite objects. He sees it as a sort of 'consciousness of being alive'[45] which itself provides a clue to consciousness of the real. On this basis, in perceiving the vague background we are nevertheless unwittingly aware of the separate objects that it contains. This also accounts for oscillation in clarity of perception because objects that are not apparently perceived as part of the undifferentiated background are, in fact, being unconsciously perceived. Our belief that we are sometimes in error about what we perceive may, in this way, be a mistake.

9. The *C* series must account for the way that particular objects of our experience appear at different times to be more or less clear to our consciousness.

We can assume that our conscious perception of objects which appears to increase or decrease (in that sometimes we are and sometimes we are not conscious of them) is, in fact, only rising or falling (in that although we are not consciously perceiving them still we are perceiving them). Our perception level oscillates above or below the level of our conscious perception that claims clarity in its perception of separate objects. If there appears no rise or fall, or that consciousness seems to have ceased, then perception of a 'conscious background' (although more usually we might consider this to be 'unconscious') still is taking place. Such oscillation can be in the observer or the observed. For example, an object may disappear from my vision or I may lose interest in an object that I have been perceiving but in both cases I may continue to perceive them. In the same way, I do not have to be present in order for the object of my love to be aware of my love for him.

If an observer perceives two different objects, *G!H* and *G!K* then the characteristics of each of these will depend in part on *H* and *K* being different. As each of *G!H* and *G!K* pass to different stages in the *C* series they will have characteristics that are different. This may lead to the 'internal harmony' of one being different from another. This would then lead to one being perceived as say, changed and the other as say, unchanged. This can bring about the false assumption that sometimes the changes occur in the objects perceived. By the same account, a corresponding change will occur in the nature of *G* and subsequent oscillations in the clarity of objects perceived by *G*. As well as this, the nature of *G!H* (as all the relations that *G* and *H* are involved in) passes along the *C* series and this would account for the possibility that the clarity of our perception of an object is directly influenced by another substance. Bearing in mind, that in McTaggart's system such oscillations appear in events because all perceptions in the pre-final stages of the *C* series are misperceived as events in the *B* series.[46]

However, only if we assume that all objects of perception persist (as selves) can we conclude that one object is perceived as changed and the other as unchanged. It is possible that an object ceases to exist and is therefore no longer subject to change. Or that change in itself means that the 'prior to change' object is different from the object into which it has changed. Just because in the *C* series no self can cease to exist does not mean that an alternate ideal 'self' could not. Only in the timeless 'eternality' of the *C* series can this be so.

10. The *C* series must account for the fact that the accuracy of our knowledge changes in that sometimes we believe a thing to be true (and it may or may not be) and at another time we do not (and it may or may not be).

It seems self-evident that there is an oscillation in the clarity of our judgement although such changes presumably could be true changes in judgemental conclusions based on changing causes that lead to that judgement being made. However, assuming that clarity of judgement does alter, if all knowledge is really perception, and perception can oscillate in its clarity, things that appear as perceptions are really either true or false judgements.[47]

11. Problems of oscillation in perception cannot be answered by a denial of the asymmetrical passage of time (which is not real). However, there can be no acceptable answer that is not connected to the apparent place of perceptions in a time series. The *C* series is a causally connected series, but it cannot be explained by causality because causality can be symmetrical and the *C* series must reflect the asymmetry of time.

In McTaggart's system, all that appears to occupy a place in the *B* series is really occupying a place in the *C* series. Any term that is placed in the *C* series will modify its content as it perceives more or is itself perceived. Although sometimes these modifications will not be always in the same direction as our perceptions of extent, clarity or accuracy,[48] the denial of time in itself cannot answer the problems associated with our experience of it. Acceptance of the *C*-system is also an acceptance of something timeless that, until the last term of the series, will always appear in time.

12. The *D* series accounts for the apparent fact that terms that are all parts of a correct perception (that is both correct and incorrect perceptions) cannot be in the same set of its parts.

According to McTaggart's *D*-system, the difference between a misperception and a correct perception does not lead to another state of misperception.[49] There is no ground in between the two states. It is (understandably) like Hegel's idea of purging erroneous cognition in the process of synthesis towards the Absolute. All terms must have some content in common. If this were not so then any difference between any terms and the whole (*G!H*) would automatically include either the term or the whole itself.[50] We have seen that McTaggart provides an acceptable explanation why the terms of the misperception series, although parts of the correct perception G!H, do not constitute a set of its parts.[51]

These twelve points (admittedly formed by McTaggart for his own positive ends but at the same time all relevant tests of parts of the theory) are all, to some degree or another, compatible with the whole system. It seems to me reasonable to conclude that the description McTaggart provides of the atemporal *C* series fits in with his fuller description of the nature of existence in general. The point being that the *C* series is part of a complete metaphysical description of reality and is not meant simply to replace time. This makes the *C* series a serious contender to fill the gap left the inadequacy of time to answer implicit problems with temporality. Any (even partial) acceptance of the atemporal system places great strain on any trust we may have that our *B*-experience reflects reality.

6.4 The C series can illuminate our ordinary conception of time

Nevertheless, we may think that the *C* series still is just an alternate description for what we think of as time. I now move on to describe how this is not so and that positively it can illuminate our puzzled view of time and serve as a psychologically acceptable replacement. I shall do this by lining up the general nature of the 'inclusion series' with the asymmetrical nature of time (§ 6.4.1) and then specifically by showing that its directionality is compatible with the apparent direction of the *B* series (§ 6.4.2).

6.4.1 The inclusion series

The atemporal series may be a reasonable (theoretical) alternative to the (apparently obvious) temporal series but, to be psychologically acceptable, it must be shown that it has some realistic, fundamental sense evident in our ordinary conception of time — metaphysical theories have little appeal if they do not fit with common sense.

McTaggart thinks that every series has two opposite senses. He thinks this view has more 'ordinary language' (an expression which includes the everyday nature of common sense but at the same time excludes the inaccessibility of symbolic logic and mathematical devices) appeal than the mathematical view that two series have the same order of terms. The first relation of the one sense (at one end) 'generates' the series in that sense and the first relation of the other sense (at the other end) 'generates' the series in that (converse) sense. If *P* has a generating relation to *Q*, *Q* to *R*, *R* to *S* and so on then *S* has a generating relation to *R*, *R* to *Q*, *Q* to *P* and so on. In the same way, the *B* series has two senses based upon the generating relations of 'earlier than' and 'later than' and the *C* series has two senses based upon the generating relations of 'included in' and 'inclusive of'. The problem, however, is to match these two series which, if not done in some way, leaves the connection too loose to be meaningful.

The two converse generating relations are inseparable and in the sense that each depends upon the other (without the one there could not be the other), equally important. McTaggart does not consider that the converse of a generating relation may not be a generating relation at all, though his conclusion that one is 'less important' than the other means that this thought is implicit.

According to McTaggart, there is a sense in which one may be 'more important' than the other. 'More important' in that the process *P* to *Q* to *R* to *S* may 'express the nature of the series more adequately' than the process *S* to *R* to *Q* to *P*.[52] If the process *P* to *Q* to *R* to *S* in the *B* series is more important than *S* to *R* to *Q* to *P*, then it should be possible to associate that more important sense of the *B* series with the inherently more important sense of the *C* series which is directed towards the final stage. Ignoring the fact that we only experience time in one direction (because this experience is erroneous), the more important sense of the *B* series (although the series can be taken in either direction) is the series of change that goes in one direction and not the other. If the *B* series is the appearance of the *C* series, then there should be some correspondence between the more important sense of the *B* series and the more important sense of the *C* series. If this can be shown, then the 'earlier than' to 'later than' relation is no longer confined to the purely temporal.

As we know from experience, the 'apparently' temporal *B* series is a series of change. Although we could take the terms of the *B* series in either direction, earlier to later or later to earlier, there is only one 'acceptable' direction of change and that is from earlier to later. If change is described in this sense (of earlier to later) then 'earlier than' and not 'later then' is more important, inasmuch as it will always precede 'later than' and is fundamental to the occurrence of 'later than'. If *P* is our starting point, then it is earlier than *Q* that is earlier than *R* and so on and therefore more important than both of them. On the other hand, if we imagine that 'later than' was the most important sense of the series, *P* would be later than *O*, which would be later than *N* and so on and none of these terms would be as important as any subsequent (possible) terms. It is only in the first example, where the 'generating relation' of importance is 'earlier than', that we get an appropriate sense of a series in accordance with the apparent direction of change. That is, that terms in the *B* series are successively present, being first earlier terms in the present and then later terms in the present. The earlier terms being the important indicators upon which we judge both the nature and occurrence of change. If, on the other hand, we took the view that we would judge the nature or occurrence of change on later (possible) events we would never be in a position to make any judgement that change had occurred or what it resembled — I cannot judge the changes that occur in respect of my growing older in relation to my condition at death.

On this basis, McTaggart thinks it reasonable to conclude that the fundamental sense of the *B* series is from earlier to later.[53] He realises that our inclination to believe that earlier determines later in a way in which later does not determine earlier, may be an erroneous pre-disposition to causality. But he finds no reason to believe that causality is universal. Indeed, he thinks that it can only be shown to extend to qualities of secondary grade parts that constitute a sufficient description of a substance.[54]

If we are troubled by disregarding causality in such an easy fashion, we can find plenty to reinforce the direction of change in common sense. Common sense makes it difficult to defend the idea that somehow in time later occurs 'before' earlier. Or that 'before' there could be anything earlier than something else, it would have to be preceded by the something that was later than it. Arguments for reversed time or backward causation do nothing to alter our experience of this and little to cause us to accept it. Arguments for backward causation do little to support an alternative view as even a series that is backwardly generated has some sense of 'earlier than' and 'later than'. Likewise, time reversal does not alter this nor does the notion that causes follow effects, for in both cases the experience of directionality happens in the *B* series order from earlier to later. If we were living in reversed time our *B*-experience would still be from earlier to later; we would simply be witnessing a later event as being the cause of an earlier event. If we were actually living backwards in time our experience would still be in the *B* series and any confusions we have here are to do with being mixed up about the order of experience and our experience of order.

If the *B* series has a direction, and this is a fundamental part of it, then because it is a reflection of the *C* series there is a strong chance that the *C* series shares a similar fundamental sense. According to McTaggart we can match this relation directly with the relations of the atemporal *C* series. The different senses of the *C* series 'included in' and 'inclusive of' differ, not only by virtue of their position in the series, but also because of qualitative difference. Because the senses of the *B* series must in some way correspond to the senses of the *C* series, he assumes that there must also be some qualitative difference between 'earlier than' and 'later than'. Indeed, this is important in his ontological quest, for if there were a variable amount of good and evil (according to whether the position is earlier or later in the *B* series) then this, he thinks, would be 'a matter of great practical importance'.[55]

There is an unusual imbalance that derives from our perception of earlier and later and our emotional responses to that perception. Anticipation generally affects our present state of happiness much more than memory does in that we can become genuinely more excited about a future event than about a past one. Even though anticipation can be dictated by past events, and this alters this asymmetry, anticipation seems the directionally guiding principle. It must be said, however, that it is much easier to form a clear picture of something that 'I have experienced' than of something that 'I am yet to experience' and this could be construed as a directional sense. But our anticipation is drawn from expectation of change and change (actual or McTaggartian) can only occur from earlier to later.

McTaggart believes that no matter how influential earlier events may be, it is those that will come later that bring us greater happiness or unhappiness through anticipation of them in the present. This is not affected by time being unreal — anticipation of good or evil in an apparent past or an apparent future still is anticipation. He is unable to find a reason the future should be so important in the present but assumes it as a fact, if not an ultimate fact.[56] Indeed, he thinks that 'future' is more important than 'later than'. For example, if two events are in the future and two events were in the past, we think of the events in the future as more important than the events in the past. This is so despite which was the earlier or the later, or which was good or evil in each pair.[57] This not commonly made distinction further reinforces the over-arching importance of our *B*-experience and the way that *A*-ordering is subsumed by it.

I think our temporal view of the past as 'dead and gone' also has something to do with us placing it second in importance to the future. This view may be quite simply a function of the tautology that 'the past is the past', but we do not mean the same when we say 'the future is the future'. In thinking of the past like this we are deeming it irrecoverable and no longer accessible to experience. That 'the future is the future', though this may imply some sort of determination or inaccessibility for alteration, it predominantly means that whatever it is (whether determined or not) is yet to be experienced. As our life is made up of experiences in the present we therefore depend on the more important sense of future in order for those experiences to happen. The past may have influenced the present but it is not an influence *in* the present. The future, even though itself not experienceable, will influence the present and the past. An event that we know we cannot experience (such as being never able to experience again a past experience) cannot rate as highly as one that we think we stand a strong chance of experiencing. Such options, even when de-temporalized, remain in the same order of priority. We will always favour the bag of sweets that we can have over the bag of sweets that we have had. This importance pivots around the present so that 'later than' and 'earlier than' are not for example later or earlier future events or later or earlier past events but events which are earlier than present or later than present.

According to McTaggart, if we accept this priority of future over past, we can see that 'how far' things are in the future or 'how far' they are in the past, has an additional effect. Some things are further in the future than others. These later future events have more importance than earlier future events. This is because the more 'points' at which it can be future lessens the number of 'points' at which it can be past. According to McTaggart, it follows from this that, 'If a certain quality, of a good or evil nature, is true of all states later than a certain state, then some states which possess this quality will be always future or present, and they will never all be in the past'.[58] In other words, the further in the future it is, the further away from the past it is until it is so far in the future that any pastness which was relevant to it becomes so comparatively small as to be insignificant. So, the 'lessening' of the number of 'points' at which something can be past is purely relative. This is McTaggart's way of accommodating obvious evil; it will not disappear from the system but in the final stage it will be irrelevant when compared to the goodness of the whole.

Though, it must be said, this strictly serial view does not sit so comfortably with our psychological view where things in the near future generally get most of our attention. But this is because 'anticipation' of future events in our own lives is different from the far-reaching 'anticipation' of the concluding future event in the lifetime of a self. Nevertheless, events after our own death cannot have the same sort of priority based on anticipation because earlier and later future events have no meaning without a specious present and anticipation can only occur in the conscious present. We must therefore make a distinction here. Although it is reasonable to conclude that later is more important than earlier, this is confined to our anticipation in the specious present and the way that the anticipation itself has an effect upon the present. The notion that distant future events are more important is meaningful only if we are looking (like McTaggart) to the ultimate event being attainment of the Absolute at the final stage of the *C* series.

So, to find a meaningful relation between temporality and atemporality it is essential that the fundamental sense of the *C* series lines up with that of the *B* series; 'inclusive of' must appear as 'later than' and 'included in' must appear as 'earlier than'. If this is not so then the final term of the all-inclusive *C* series (that has a value that infinitely exceeds all other terms) would be at odds with the appearance of the *B* series. Such a difference would cause the *C* series to be too unalike the misperceived *B* series to be credible as the reality that lies behind time and, as McTaggart points out, 'the fact that it is desirable does not make it more likely to be true'.[59]

Nevertheless, if the *B* series has a fundamental sense then, even if it is completely misperceived, anything that appears fundamental in the *B* series must correspond to something that is fundamental in the *C* series (if the *C* series is real). In this way they are like identical images on two transparent sheets. When laid one on the other, the images must correspond. The fact that one sheet is red and the other blue does not affect their fundamental identity but only the shade by which we can conveniently recognise their superficial difference.

6.4.2 Directionality

One sense in which the *B* series is fundamental is because of the 'direction' of change and yet change in McTaggart's theory is a misperception. This seems like red being both red and blue simultaneously. If 'direction' of change is fundamental to the *B* series, there must be some way of understanding it in relation to the atemporal series.

There are certain characteristics of the *C* series that we can assume are manifested as change in the *B* series. For example, the fact that in the *B* series *N* succeeds *M* and *O* succeeds *N*, is a manifestation of the fact that in the *C* series *N* comes between *M* and *O*. That this shows itself as change in the *B* series does not mean that it is change in the *C* series. Indeed, in McTaggart's system, such temporal change is identified as an atemporal serial increment.

But there *is* a connection between the world of error and the reality that lies behind it; correct perception and misperception both exist as perception, and misperceptions that we have in the *B* series are at least of something *appearing* to us in time. Because the *A* and *B* series depend upon the *C* series and the *C* series is real, then the misperception that we experience in time must also be an experience of something real. From this we can assume that the asymmetrical nature of our *B*-experience is reflecting something of the serially successive nature of atemporality in the *C* series. If this is so, then there must be a link between what appear as change and the direction of change on the one hand and their atemporal counterparts on the other. Because time is not real, we need to modify our view of 'actual' change as it appears in time. Only by using an extended system of 'real' change can we begin to see beyond the errors of matter and time. *Cambridge change* provides such a system whereby every change in any event changes every other event in the universe.

Our common understanding of 'actual' change only seems to prove a temporal world. But this proof is invalid because 'real' *Cambridge change*, unlike 'actual' change, is not a reflection of things changing in time but a reflection of the seriality of the *C* series and is established by the addition of the descriptions of events. As time is a misperception so 'actual' change, as we conventionally determine it, is also a misperception. Any idea that we have of actual, material change in things is mistaken because McTaggart's universe does not contain any material things, and any idea we have of things changing in the *A* series is mistaken because the *A* series is self-contradictory. If things, and facts about them, do not change, then things and facts have a permanent, unchanging position in the *B* series. Consequently, our experience of time, based upon things changing, cannot truly be reflecting something temporal because temporality requires change to be taking place. When we think that some 'thing', or some 'fact about a thing', has changed, in truth it is different events that are attributable to that thing, or fact about it, that have changed.

In order for 'atemporal change' to happen, *Cambridge change* must be possible without the *A* series. On the face of it, it seems unlikely that *Cambridge change* can take place in the permanent *B* series any more than actual change in things or facts about them; after all, any event that was earlier than another event will always be earlier than that other event. For example, the event of the death of Queen Anne can seemingly only change in relation to the terms of the *A* series; that once it was in the future, then it was in the present, then it was in the past. Without the *A* series, the event of her death and any associated events (the causes of the death, its effects, that it is a death, that it is the death of Queen Anne) all remain permanently fixed in the *B* series just as much as Queen Anne herself.

But there is a way out of this. There is actual change when I am able to assert one identical proposition which is at one time true and at another false. The statement 'Queen Anne is alive' which is true at time t but false at time t^1 (Queen Anne having died) refers to actual change in Queen Anne. If, on the other hand, at time t we make the true statement 'Queen Anne is more alive than her father' and at time t^1 make the false statement 'Queen Anne is more alive than her father' (Queen Anne being dead) we can draw a different conclusion and gain a different view of change. Although Queen Anne still has undergone an actual change, in *Cambridge change*, so has her father, for whereas once he was 'more dead' than his daughter, now he is 'equally as dead' as she is. In this way, change is no longer only a difference in facts about things in time, but a means of establishing the relatively different states of events at different positions in serial succession.

Changes in events that occur in *Cambridge change* are not reliant upon a *B* series because it is the difference in inclusiveness of related events that indicates change as opposed simply to their different place in the succession from earlier to later. This means that in *Cambridge change*, no matter how seemingly insignificant or tenuously related, any new event is a change and any change in any connected event changes the relative status of all other events in the universe. My thinking about the death of Queen Anne now, changes the event of her death even though I have no direct access to when the central feature of her death (her dying) occurred. Because her dying is only part of the extended system of change which is the event of her death, the event of her death is no longer in the permanent asymmetry of the *B* series but in the one-dimensional seriality of the atemporal *C* series — the concept of events being earlier or later is subsumed by them being more importantly 'included in'. In this way, I can keep adding, and so changing, events related to the death of Queen Anne. As long as I do not become confused by thinking that the event of her death is confined by the actual occurrence of her dying, the idea of change in an atemporal series makes sense.

Change interpreted in this way links directly with the *C* series because the 'changing' of events is simply a serial inclusion. *Cambridge change* subsumes actual change and does not require the *A* series. For example, we would commonly assume that when Dr Jekyll becomes Mr Hyde he 'changes' into Mr Hyde. This assumption leads us to believe that not only does this change require an *A* series because it happened in the past, but that the present situation of Dr Jekyll now being Mr Hyde represents reality. In *Cambridge change* neither of these assumptions is correct because Dr Jekyll constitutes part of Mr Hyde so Mr Hyde's reality is both Mr Hyde and Dr Jekyll. Dr Jekyll is not in Mr Hyde's past, he is part of what Mr Hyde is now so there was no point in time when Dr Jekyll 'changed' into Mr Hyde. Though it is convenient to judge Dr Jekyll as both different from Mr Hyde and as having existed in Mr Hyde's past it is still a superficial view of what is in reality the timeless inclusion series of Mr Hyde which includes Dr Jekyll (and all previous incarnations). We can on this basis, consider time based on actual change less fundamental than timelessness which does without it.

Nevertheless, accepting *Cambridge change*, although making actual change in the *A* series redundant, does not provide us with a 'more important' sense to the apparently different states of succession in the *B* series. *B*-states remain different even without time in the *A* series, and, if we go no further, *Cambridge change* simply infers a version of time in the *B* series which, although unlike our normal view of time, does not have to be the *C* series. *Cambridge change* only reinforces the notion of permanent seriality and still we are left to wonder how we can harmonise this with both an apparently temporal, and a not so apparently atemporal, world. Although *Cambridge change* may be acceptable as an over-arching principle for change, unless the serial view of events in the *C* series can be shown to be more convincing than the temporal view in the *B* series, we would probably tend to stick with the devil of time that we know than to look to a counter-intuitive series of atemporal succession that we cannot know. To resolve this we need to show that the *B* series has a fundamental direction — a 'more important' sense — and that this resembles the 'more important' sense of the *C* series.[60] Doing this, of course, exposes the inconsistency of, on the one hand, accepting that we always misperceive the *C* series as the *B* series, and on the other, using the *B* series to justify the direction of the *C* series. This seems a bad case of wanting our cake and eating it. Nevertheless, because the direction of the *B* series seems an incontrovertible part of our experience, and because *B*-terms always fit our experience of ordering, there is no way around this and if we can establish *some* equivalence in *C*-terms we will not be eating *all* our cake, indeed, biting into it may reveal it for what it really is — that is not a cake at all.

Our experience (and any interpretation we make from our experience) is always in the *B* series. Because of this, the connection that McTaggart is trying to establish between the distinct qualities of the temporal *B* series and the atemporal *C* series will always be difficult to accept. For example, we generally assume that our perceptions have a temporal order. We also assume that the story that the perceptions (themselves in a temporal order) tell us, also has a temporal order. We may (like Kant) think them distinct or we may assume them to be the same. McTaggart's self perceives things in a temporal order (the *B* series) and the story that they tell may be perceived (either separately or as the same) in a *B* series (together with an *A* series). However, this perception is a misperception. The experience of order is an increment in the *D* series of our misperception of the *C* series. The *really true* nature of our perception is of a perception of a serial order that is atemporal. In the atemporal world both of these assumed states of perceptual ordering is atemporal.

There is no compelling reason to disbelieve (after all it cannot be known) that the relation earlier to later corresponds with the serial sense of the *C* series. However, it would be useful to determine which of the terms of the series is fundamental. On the face of it, there is no indication whether 'included in' or 'inclusive of' is the fundamental sense. What is certain though, is that change cannot be a deciding factor as it is confined to the *B* series.

McTaggart suggests that the answer can be found in the sense of a series that proceeds from premisses to conclusion, each third term being inferred from the previous two. A series expressed in this way has no temporal change in it, though if contemplated in a temporal series, it will exhibit an intrinsic direction of its own from premisses to conclusion. In addition, the terms of such a series need not be discovered in this order. Armed with the conclusion we could (according to McTaggart) enquire from what premisses it could be inferred.[61] McTaggart thinks that such a series can be gone through in any order, though its overwhelming sense is from premiss to conclusion and not from conclusion to premiss. This again is to be found in Hegel's dialectic, where if the validity of the category of Pure Being is admitted, then all other categories follow from it by inference. On the face of it, we could not apply this to the *C* series because the *C* series is not a series of inferences and its terms are not propositions. However, in the *B* series (although no term is directly inferred from any other term either earlier or later) we could infer relations of terms by induction based upon general laws. McTaggart thinks that this method can be applied to the *C* series.[62] But the inference possible in the *B* series as causal cannot be applied to the *C* series to demonstrate any fundamental sense as in the *B* series there is no reason to suppose one direction of causal implication as opposed to another.

McTaggart supposes that each term can be assigned its place in the *C* series by virtue of the amount of content of the whole that is included in it. The greater the amount of content the nearer the whole, the lesser the amount of content the further away from the whole. But he realises that this would mean that the position of each term could only be defined in relation to the whole. The terms could not be defined in relation to nonentity (the other 'end' to the whole) because nonentity is the boundary of all *C* series whatever wholes they are. Therefore position in the *C* series could not be assigned according to how much its content exceeded nonentity, because it would not be known in which *C* series the term could be found.[63]

McTaggart thinks defining position in the *C* series with respect to the whole may be possible in another way. Because the whole is a term in the *determining correspondence* system it can be stated without need for reference to any other term.[64] *G!H* is a correct perception of *H* whereas all other terms in the series are misperceptions of *H*. The nature of misperception of *H*, or indeed, the fact that the perception is a misperception of *H*, can only be judged by comparison with the correct perception of *H*. Also, the correct perception of *H* can be expressed without any reference to the misperception of *H*. On this basis, he thinks we can conclude that all terms in the series stand in the relation 'included in' to the whole. Although the whole does not stand in this relation to any of the other terms, it does stand in the relation 'inclusive of' to every other term. We can therefore conclude that the fundamental sense of the series is from 'included in' to 'inclusive of'.[65] If this is so then the *C* series is limited at its beginning by nonentity and has as its final term the whole. The direction of the *C* series goes from less inclusive to more inclusive.

McTaggart thinks that this provides reason enough to believe that 'included in', in the *C* series, is what we misperceive as 'earlier then' in the *B* series. Anything that in the *B* series, happens to something later, will be included in the earlier state of that particular something and therefore become 'inclusive of'. This is therefore inkeeping with our misperception of the *B* series. We do not perceive (even if by misperception) one series with a fundamental sense that appears to be fundamental (that is the *B* series direction from earlier to later), while perceiving another series as having a fundamental sense that does not appear to be fundamental (that is a *C* series that runs from 'inclusive of' to 'included in').[66]

The practical consequences of McTaggart's conclusion that the direction of the *C* series is from 'included in' to 'inclusive of' are very important. Had it been the other way around we would be faced with a past that was infinitely more good than the future will ever be. For McTaggart this would upset his main philosophical aim. For him the whole must be in the future to give meaning to the idea of increased good and the continued existence of selves throughout a timeless universe.[67] On McTaggart's argument we are led to assume that the whole will be in the future, for *sub specie temporis* it is the last term. It cannot be perceived as present because it is a correct perception and as such it cannot be perceived in time (although it can be perceived by others as in time).

So, is it possible to increase the psychological appeal of a system that condemns us to an unavoidable experience of time and seemingly bars us even from a glimpse of what is timelessly real? I think things are not as bad as they seem. Conscious experience, our specious present, has a known place in the *B* series. Consciousness informs us about the order of events we experience. Ignoring the possibility that we could be in a stage of oscillation that rules otherwise, it would be inconsistent with accepted common sense to think that we experience 'later' before 'earlier'. Unless the *B* series has some unalterable 'standard' direction to which we need not conform, our experience of order will always be the same. It will be the *B*-succession no matter what direction time or cause is going in. It occupies a position that reveals a difference between that which *may be* experienced in later times and that which *has been* experienced in earlier times. Earlier is qualitatively different from later because it includes all that was previously earlier than the present, and therefore is something that was once experienced, whereas later, arguably, includes only the 'possibility' of that which has not yet been experienced. The sense of events which are earlier or later only in the past does not have the same qualitative difference as events which are earlier or later than the present. Though this sense has a general application to our fuller understanding of *B*-ordering and our ability to organise our memories, because it exists only in memory it is a confined (or reduced) sense of the *B* series and has more to do with the chronological placing of events in our history and less to do with our current experience of time. In this way, not only does consciousness place us central to the asymmetry of time, it also reveals the direction of our experience of events and, from this, we can imply a 'more important' sense.

Although later is less substantial, in that earlier times are more tangible because they have happened, to consciousness it is later which claims more importance. Only in later times can we take action and change events. Anticipation can only be applied to later times and we can only utilize information from our stock of earlier experiences in later times. In addition, in *Cambridge change*, subsequent events include not only events which are new but also, as currently earlier events, all previously later events. The conscious balance is therefore always tipping towards later which must, I think, give it a more important sense. This is so even if, in *A*-terms, the future is determined (as McTaggart believes). Although the future is determined in the sense that the *C* series is unerringly proceeding towards the final stage, this does not preclude individually meaningful decision-making — though the determinist's choice is determined, still he can determine events by his choice.[68] The determined future, therefore, includes the possibility of action whereas this is not so in the case of the past.

If the more important direction of the *B* series is always from earlier to later and the *C* series is different from it (because it is atemporal) yet resembles it (because it is our misinterpretation of the *C* series which gives us an experience of the *B* series), then conceivably the directionality of the *C* series is the same. After all, the fundamental features are similar and if we exclude time we gain by excluding contradiction. If this is so, then an event which is earlier (that is, temporally prior to our experience) would be atemporally 'included in' perception, and an event which is later (that is, temporally posterior to our experience) would be atemporally 'inclusive of'. Assuming that the *C* series does not possess any other relations that could appear as *B*-relations (which is unlikely as it is a simple one-dimensional series), the successive and individual nature of the terms of the *B* series are equivalent to the terms of the *C* series. Mistaking the *C* series as the *B* series is a substantial error. The *C* series is not simply an alternative way of describing the *B* series but a way of replacing our erroneous experience of time with a system of timelessness which is not only more coherent than time but also explains why our *B*-experience is incorrect.

Though this seems plausible, it might be objected (at the least) that we are placing great emphasis on the *A* series. This is because the *B* series, which we are claiming as our window to the *C* series, relies upon it, while at the same time we are assuming it to be self-contradictory. This emphasis, however, actually supports the argument. The apparent temporal movement in the *A* series is a fundamental part of our conscious experience and its self-contradictory nature does nothing to alter this. Indeed, if we accept that the *C* series is real, then it is the *A* series together with the *C* series that give us the erroneous experience of time in the *B* series. The *B* series, no matter how pervasive, is not fundamental because the *B* series is a reflection of the *C* series and it is also reliant on the *A* series. Neither the *A* nor the *B* series exist independently; it is only when we incorrectly apply, by the act of consciousness, the idea of earlier and later to the *A* series that we get an impression that time is passing. As we have already seen, the fact that something is earlier than something else, in itself, does not give us an impression of time. The permanence of the *B* series does not give it precedence over the transitory *A* series which it relies on, but, as we have seen, its permanence does make it more similar to the *C* series than the transitory *A* series. This link between the *C* series and the *B* series can only be made if the *B* series is a conscious experience and this can only be the case if it relies on the *A* series. Our emphasis on the *A* series, even though it is flawed if we were to attempt to prove the reality of time, therefore actually supports identifying our *B*-experience with the *C* series. In this way, we might say that time itself gives us good reason to believe it is not real.

I think this raises substantive issues about how we can interpret our experience of time. We have seen that actual change is only part of *Cambridge change* and *Cambridge change* offers us the possibility of interpreting changes in events without resorting to time. Also, we have seen that we can establish a more important directional sense of the *B* series and this equates with a more important directional sense of the *C* series. Although doing this utilises the *A* series, we concluded that the fundamental relationship between the *A* and *C* series and the subordination of the *B* series to the *A* series, actually supports the comparison. From this, I think, we can reasonably suppose that the *B* series can be interpreted in *C*-terms and two fundamental features of time (change and the direction of change) can be replaced with atemporal features. We are, however, still left to wonder if doing this increases the psychological appeal of timeless eternity.

I conclude that it is reasonable to accept that the inclusion series can be lined up with the asymmetrical nature of the *B* series. We have to recognise, however, that the direction of the *C* series is strongly connected with McTaggart's desire to show how the *C* series will lead to ultimate good. However, even though we may doubt the psychological appeal of replacing time with timelessness, change and the direction of change can be interpreted atemporally.

6.5 Fundamental relations between the temporal and the atemporal world

It may not be sufficient simply for us to appreciate that the *B* series can be substituted by the *C* series. If this is all we can know of the *C* series then we are led only to a pointless form of atemporal surrogacy. What we need to know is that there are enough reasonable possibilities for atemporality to replace the *B*-world as to cause the *B*-world to be more than suspect (and preferably more under suspicion than the *C*-world). Consequently, in this section I will describe some fundamental relations between the temporal world as we experience it, and the atemporal world of the *C* series. By doing this, we will be able to see that thinking that we experience the *C* series is an acceptable view to hold and places us beyond the confusions that the world of time continually presents.

So far, we have seen that all the terms of the *C* series appear as the *B* series. Our perception of the *C* series also appears as an *A* series. From our position at each stage in the *C* series all that is on one side appears as future, all that is on the other side appears as past and the stage itself appears as the present. As we proceed along the *C* series the future seems continually to become present and the present seems continually to become past. The *B* series is dependent upon the *A* series but because the *A* series is not real so the *B* series is not real. On top of this, neither is there any real change. The only reality lies behind the erroneous world of time in the *C* series.

To move from theoretical 'replacement' to a reasonable atemporal interpretation of the *B*-world we need to apply the principle of atemporality to a broader range of assumed qualities such as: the nature of the present, the idea of a common time series and the idea of infinity.

In the atemporal world the present is an overlapping series of terms of varying magnitude that may vary from self to self. Such a present (that consists of simple parts) is a duration indivisible into parts of parts to infinity. In this sense, if no simple term is next to another simple term then the number of terms may be infinite. If, however, the simple terms of the *C* series are next to each other then the duration of that present cannot be divided infinitely.[69]

This serial 'place' is a sort of specious present because it is the place of conscious perception. Indeed, we have seen that 'present' experience of the *C* series is not at odds with 'present' experience in time. There is, after all, nothing strictly temporal about perceptions when we 'collect' them. Perceptions themselves are timeless and we are merely led to think of them as temporal because we appear to collect them in time. The truth of the matter is that if we reduce any perception to its smallest possible component we also instantiate it and remove it from the temporal arena. Thinking that events occur in time is therefore a result of only being able to appreciate perceptions in collections of components as opposed to as components themselves. The *C* series does not suffer from this problem as each stage is defined by the smallest unit of perception (whatever that may be). We can, therefore, accept the idea that serial, atemporal placement is the same thing we experience as temporal presentness (although the latter contains probably a greater number of atemporal places). If we accept that our experience of perceptions in time is a convenient way of 'packaging' perceptions as opposed to the way perceptions actually are, then a replacement system that equates with many of our (confusingly experienced) experiences, but is atemporal, causes no real psychological problems.

We have already seen that there are reasons why we should think that there is a common time series. All perceptions fall within a self, all perceptions of time are misperceptions and no two selves can have a misperception in common. McTaggart says that this is no more likely than two individuals having the same fit of toothache.[70] Selves may have similar misperception (presumably like both having toothache) but they cannot have the same misperception. On this basis, it would be difficult to accept that our apparently common experience of time is in any way true. When we imagine 'time' somehow existing separately from our experience, we think that it has common qualities that we can all share. Once we recognise that time is not a thing-in-itself, but relies upon each individual's experience of it, such commonality is lost. Like space-time theory, the atemporal world has no common series. Each *C* series is a series in the perceiver and as such does not depart (unless we hold a genuinely absolutist view) from the idea of 'unshared' time that is not strictly common to us all. Because the self's perceptions continually cross-over with perceptions had of it by others there is an understandable (but erroneous) conclusion that perceptions are either occurring simultaneously or within some sort of mutually shared time-frame. But this is not a correct conclusion. If someone tells me that they have toothache and I say, 'I feel for you', I do not mean 'I feel it'. Added to which, by the time they receive my expression of sympathy the toothache may have passed.

Even if there were a real time series this does not necessitate common time. Two perceptions (or misperceptions) could be differently fixed in the time series as present or at a point in the past as earlier or later than another point in the past. Admittedly, events in apparent time are the same as the order in the *C* series but the *C* series provides a superior description of their ordering than does the confusing system of time. The appearance of time, is what McTaggart calls (borrowing the term from Leibniz), a *phenomenon bene fundatum* but its inescapability does not make it real. The only reality to time is its inescapability as an error of judgement.

If the appearance of time is therefore an inescapable phenomenon, then a common time-series between different selves in the *C* series must also be a *phenomenon bene fundatum*. For there to be a common 'time' series (in the *C* series) there must be some sort of common simultaneity, as well as a real sense of earlier and later. If there was only one 'other self' to perceive that would be straightforward. If *H* and *G* both perceive *L*, then the perceptions that *H* and *G* have of *L* in any part of the *C* series of *L* as present would be simultaneous. *H* and *G* are in a simultaneous relation to *L*. If this were so then there is evidently a common time-series. But this is not like the reality of the *C* series that, by *determining correspondence*, consists of selves each of whom perceives more than one self. The best that can be said for there being a common time-series in the *C* series, is that although *G* and *H* cannot have the same inclusion series, they can have a 'correspondent' inclusion series. This would provide, like the time-series, for each self a common time-series *phenomenon bene fundatum*.[71] This, however, would be a coincidence of correspondence because of the number of selves and not a fundamental quality of the nature of the *C* series itself or the experience of any self within it.

We can imagine time to be finite or infinite. There is nothing in our experience of time that tells us which it could be. In the *C* series, in one direction, there is (most probably) a last term that is the fully inclusive term that contains all its parts. In the other direction the limit is nonentity. This final stage however, does not end — it cannot as there is no time for it to end in. It appears in time as endless, though in fact it is 'eternal yet timeless'. Indeed, all the pre-final stages in the *C* series are eternal but again they do not appear as endless in time. The last stage of the *C* series is endless not because it appears as endless but because there is simply no stage beyond it.[72] If the atemporal system were infinite, it should also be infinitely divisible. This would be so if it only had simple parts as its terms, or simple parts with no next terms.[73] But parts in the *C* series cannot be divided into parts unless those parts are determined by *determining correspondence*. As the terms of the *C* series cannot be determined by *determining correspondence*,[74] the terms of the *C* series cannot be divided into parts of parts to infinity. Of course, if the *C* series cannot be divided in such a way (assuming that it represents reality), then neither can the time series. In other words, the *C* series cannot contain infinite divisibility though it could be an infinite (one-dimensional) series.

Whether the terms of the *C* series consist of parts that somehow are not next to each other, and are therefore infinite in number, seems indeterminable.[75] McTaggart sees no logical contradiction in the series having no next terms. The fact that the *C* series appears as the *B* series does not alter this. The *C* series must have at least as many parts as appear to us as parts in the *B* series, but that number, which we perceive as the *B* series, is always finite. However, the *C* series still may be infinitely divisible, because it may contain smaller divisions than we can perceive in the temporal *B* series.[76] If the *C* series is infinite then the Absolute will never be attained, so McTaggart considers it finite. He thinks the universe is 'very large' and believes the final stage will be attained only after much oscillation in, perhaps seemingly unending, time.[77]

This reinforces our conclusion that the *B* series is an inescapable phenomenon but there is a strong argument, particularly in respect of the notion of a common time series, to uphold the acceptability of the real world as atemporal.

6.6 Actual and apparent time

Further to consolidate on the division between that which we inescapably experience and that which really is, in this section I look more closely at the distinction between apparent and actual time. I think it is useful to call both apparent and actual time '*true* time' (the temporal order as we experience it) and that what lies 'beyond' this erroneous experience the '*really true*' 'time' of the *C* series (the atemporal order that in McTaggart's idealist world it really is). In this way we can keep the confusing notion of our unavoidable experience of actual time as *true* time, separate from a plausible explanation of what it may really be in *really true* 'time'. Further, if our experience of time is in fact an experience of an underlying reality then this classification enables us to distinguish between our '*actual*' *C*-experience and our '*apparent*' *B*-experience of time.

McTaggart points out that there is a genuine psychological difference between periods of time filled with exciting events, which in retrospect seem longer, and periods of boredom, which in retrospect seem shorter.[78] If we are sceptical about the nature of apparent time, this could lead us to suppose that such a difference is a real variation in lengths of time. However, it seems more likely that during periods of tediousness we pay more attention to time in a short period, than we would otherwise pay to it in a long period. Conversely when well occupied by interesting experiences, we are so occupied by our interest that we pay scant regard to the elapse of time, and so conclude that little time has elapsed. Even so, a period filled with exciting events may lead us back to the conclusion that this greater number of events must have been spread over a greater amount of time.

The important aspect of this is that the effect is produced in retrospect. Our experience of any one specious present does not seem longer or shorter than any other at the actual time at which it occurs. Though, as we have already seen, our experience of the specious present is itself subject to differences (the dying man's 'life flashing before him' may be an example of this). Nevertheless, the judgement that we are making about 'how long time is' (no matter how we reach this conclusion) is a mixture of memory and present experience. There is, strictly speaking, no way of knowing 'how fast' time is going now, only 'how fast' it seems to have been going in a previous present experience evaluated from my current present position.

If time is in fact unreal then in truth there can only be the appearance of time. This is not a contradiction. It means that time cannot be a fundamental reality because the conflicting differences of duration by appearance, as McTaggart points out, do not have the uniform characteristics of the *C* series.[79] Any stretches in the *C* series that have equal proportion, or any given proportion one to another, will appear as equal or proportionate stretches in the *B* series. The appearance that stretches in the *B* series may be equal or proportionate is therefore not erroneous. Though if we conclude that they are equal or proportionate periods of time, we would be in error.

Added to this, the appearance that a stretch in the *C* series that is equal to another stretch in the *C* series, appears in the *B* series as a longer stretch (say to someone bored than to someone excited) is 'doubly erroneous'. It represents periods in the *C* series as in time, which they are not, and it represents them as unequal, which they are not.

To keep this distinction clear, I think it is necessary to modify our understanding of 'actual' and 'apparent'. The common sense distinctions between 'actual' and 'apparent' time still can be retained, even if (as in McTaggart's system) it is truly the case that all time is appearance. When we talk of *actual* or *apparent* time in both cases we are talking about *true* time as we experience it. Both *actual* and *apparent* time can be classified as *true* time because in everyday terms that is what both are. However, in fact neither the *actual* nor the *apparent* nature of time in *true* time is *really true*. The *really true* nature is the nature of the atemporal *C* series. As far as our everyday experience of time is concerned, *true* time is acceptable. In the same way that relativity theory accepts that for all normal (sub-light speed) purposes everyday time is all we need. So, if we are not concerned with the true nature of existence (and merely want to tell the time and get a picture of how time passes) then *true* time will do. However, our *true* experience is an erroneous experience of the *really true* atemporal *C* series and if we are concerned to know what really is then *true* time needs to be thrown out.

Plato gives us a clear view of this type of distinction in his analogy of the divided line.[80] This is further amplified in his analogy of the cave.[81] I am here considering both categories of the *true* to be roughly equivalent to the lower segment below Plato's line that, for him, represents the visible world. The *really true* is loosely equivalent to the upper segment that for Plato represent the intelligible world. The lower segment is concerned with reasoning about objects in the visible world. The upper segment puts this reasoning to use in its discovery, by intuitive knowledge, of the Forms themselves. For a three part division, perhaps more along the lines I am proposing here, we can turn to Spinoza. His way to the liberty of the *really true* is governed by the way in which our intellectual powers are used. The lowest level of 'vague experience' is knowledge via the senses. The second level of 'adequate ideas' is based on the general ideas that are commonly accepted. The third level is 'intuitive knowledge', which although based on the second level, reaches beyond it to a knowledge of the essence of things.[82] In my division, *true* time equates with Bergson's division of time. *Actual* time is like Bergson's *le temps* and *apparent* time is equivalent to our experience of time, what he calls *la durée*. *Actual* time is therefore the measurable time of science and *apparent* time is our subjective experience of its passage. What I am proposing here, is the notion that both *actual* and *apparent* time are only (erroneous) images of what is really going on and what is really going on is not temporal, even though until the system ends it will always appear so.

I conclude that we should alter the common distinction between *actual* and *apparent* time. 'Actual' should not be used to denote time which is real as opposed to 'apparent' which is only an impression that is somehow caused by time. Because of the unavoidability of our experience of time, both 'actual' and 'apparent' can be called 'true'. Calling them 'true', however, does not make them real. If we want to know what really is then we must look behind time to what we can call 'really true' ('time') which is the atemporal *C* series (and, of course, not really time at all).

6.7 Conclusion

In § 6.2 we saw that the *C* series allows us to interpret our experience in a way that the puzzling nature of time constantly confounds. We also saw, in § 6.3, that the *C* series forms a coherent part of McTaggart's complete metaphysical description of reality. These two features move us forward from the 'sticky ground' of arguing about McTaggart's paradox. The first shows us that we do not have to accept that time is 'by its nature' puzzling. The second shows us that the claim for the atemporal world is not simply a way of getting around the apparently logically contradictory nature of the *A* series, but forms part of a well-worked metaphysical exposition. Added to this, in § 6.4, we saw that the *C* series can illuminate our ordinary conception of time. It was seen that the 'inclusion series' can be lined up with the asymmetrical nature of time and that its directionality is similar to the apparent direction of the *B* series. This re-enforces our acceptance of the atemporal series because it provides a credible, atemporal explanation for our confusing experience of the *B* series while at the same time keeping us in touch with the unavoidably temporal world. Although, in § 6.5, we saw that time is an inescapable phenomenon, I concluded that still we can hold to the notion of the atemporal series and look for fundamental relations between it and our inescapable experience of time. After all, because the phenomenon is inescapable does not make it any more (or less than) phenomenal. In § 6.6 I categorized *actual* and *apparent* time as both parts of *true* time and the atemporal *C* series as *really true* 'time'. My belief is that we need to extend our view of time in this way if we are to understand the *C* series and its relation to the temporal world as we experience it. I think this is an important distinction and one we can use for practical purposes. For example, knowing that we are in a serially building world that retains all its contents when it reaches the final stage, as opposed to living in a fleeting world of time where everything is 'lost' in the past, provides us with an entirely different basis upon which to form beliefs about the meaning of our own existence.

I conclude that it is reasonable to look beyond McTaggart's claim that time is unreal. However, I acknowledge that his atemporal world is not easy to accept in total. Indeed, this is at the root of the problem. Because McTaggart proposes a complete metaphysics we get easily bound up with his ontological quest. This can lead us to reject the atemporal world for the wrong reasons. To remedy this we need clearly to separate the time of our experience from atemporality which is the reality that lies behind it.

Notes

1 McTaggart, *The Nature of Existence*, § 526.
2 Ibid., § 527.
3 Ibid.
4 Ibid.
5 Ibid., § 537-9.
6 Ibid., § 543.
7 Ibid., § 544.
8 Ibid., § 419.
9 Ibid., § 550.
10 Ibid., § 551.
11 Ibid., § 515.
12 Geach, *Truth, Love and Immortality*, p. 103.
13 McTaggart, *The Nature of Existence*, § 554.
14 Ibid., § 124.
15 Ibid., § 558.
16 Ibid., § 566.
17 Ibid., §§ 724 and 567.

§ 567 is the end of Draft *C*. It is from here that Broad continues his editorial effort working from Draft *B*. McTaggart's intention, as with all of his work for publication, would have been to complete five drafts. Draft *B*, a handwritten copy, has many notes and amendments.

When Geach was undertaking his research for *Truth, Love and Immortality*, he only had available the MS for Draft *A* and notes for Draft *B* of *The Nature of Existence*, vol. 2. He is now aware that a typewritten copy of Draft *B* of Chapters 32 and 34 to 42, and the handwritten Draft *C* of Chapters 32 to 43 are available at Trinity College Library, Cambridge. This means that except for only several Chapters, the MSS and TSS available are substantially those available to Broad. Had this material been available to Geach he may not have experienced problems reconciling Chapter 34 from Draft *A* to the printed text and may not have arrived at the same conclusions about Broad lacking sympathy with the printed text (Geach, *Truth, Love and Immortality*, pp. 85-8).

18 McTaggart, *The Nature of Existence*, § 567.
19 Ibid., § 568.
20 Ibid., § 571.
21 Ibid., § 531.
22 Ibid., § 572.
23 Ibid., § 574.

24 Ibid., § 579.

25 Geach, *Truth, Love and Immortality*, p. 152.

26 Geach describes McTaggart's use of *G!H* and c_x*G!H* (in §§ 579-80) as a 'slip of the pen for "*G*" and "c_x*G*"'. (Ibid., 152). Geach's further statement (Ibid) regarding McTaggart's usage of the term c_x*G!H* (*The Nature of Existence* §§ 579-80) is perhaps Geach's own 'slip of the pen'. Although McTaggart is talking about *fragmentary parts* which this symbol represents, he does not use it here.

27 Geach, *Truth, Love and Immortality*, p. 151.

28 McTaggart, *The Nature of Existence*, § 589.

29 Ibid., § 582.

30 Ibid., §§ 614-15.

31 Ibid., § 586.

32 Ibid., footnote to § 587.

33 Ibid., § 588.

34 Ibid., §§ 520-36.

35 Geach, *Truth, Love and Immortality*, p. 150.

36 McTaggart, *The Nature of Existence*, §§ 622-88.

37 Ibid., § 593.

38 Ibid., § 594.

39 Ibid., §§ 595-98.

40 Ibid., § 598.

41 Ibid., § 599.

42 Ibid., § 600.

43 Ibid., § 602.

44 Ibid., § 603.

45 Ibid., § 604.

46 Ibid., § 606.

47 Ibid., §§ 622-78.

48 Ibid., § 608.

49 Ibid., § 565.

50 Ibid., § 566.

51 Ibid., § 609.

52 Ibid., § 697.

53 Ibid., § 698.

54 Ibid., § 218.

55 Ibid., § 701.

56 Ibid., § 703.

57 Ibid., § 705.
58 Ibid., § 706.
59 Ibid., § 707.
60 McTaggart argues that the direction of the *B* series is fundamental (§§ 694-709) and that this implies the order of the *C* series from 'included in' to 'inclusive of' (§§ 710-26). He introduces a mixture of considerations to support his claim of the fundamental sense of the *C* series and its corresponding relation to the *B* series for example: chains of inference and categories of the dialectic (§§ 713-16), causation (§ 716), how the terms of the *C* series relate to the whole (§§ 717-22), the inadmissibility of errors in perception (§ 724) and our inclination to believe in the results we favour (§ 725). Geach thinks this a 'complete failure', supposing that the idea of a fundamental direction of the *B* series is merely that '...*earlier* events appear to be present *earlier* than *later* ones' and that the opposite is just as reasonable '...because *later* events appear to be present *later* than *earlier* ones' (Geach, *Truth, Love and Immortality*, p. 159). I think Geach is wrong to assume this. Manipulating the relations of the *B* series in this way introduces an impermanence into the *B* series which conflicts with any experience that we have of ordering. I agree with Geach that McTaggart's subsidiary argument, that our anticipation of future affects us more than our memory of the past (§ 701), has more strength (Geach, *Truth, Love and Immortality*, pp. 159-60).
61 McTaggart, *The Nature of Existence*, § 713.
62 Ibid., § 715.
63 Ibid., § 718.
64 Ibid., §§ 718-20.
65 Ibid., § 721.
66 Ibid., § 724.
67 Ibid., § 725.
68 McTaggart, *Some Dogmas of Religion*, § 139.
69 McTaggart, *The Nature of Existence*, § 611.
70 Ibid., § 613.
71 Ibid., § 614.
72 Ibid., § 745.
73 Ibid., § 621.
74 Ibid., §§ 537-40.
75 Ibid., § 621.
76 Ibid.
77 Ibid., § 913.

78 Ibid., § 617.

79 Ibid.

80 Plato, *Republic*, in *The Dialogues of Plato*, with Analyses and Introductions and trans. by B. Jowett (Oxford, Clarendon Press, 1875), pp. 1-519, Bk. 6.

81 Ibid., Bk. 7.

82 See Spinoza, *Ethics*.

7 Conclusion

In Chapter 2 I made the assumption that it was necessary to introduce McTaggart. I did this by describing his work, some of his terms and some important features of his fuller philosophy. We saw from this that his denunciation of time, though it can be properly judged as an argument in its own right, has only limited meaning if not placed in the context of his fuller metaphysical exposition. McTaggart's argument against time is not an isolated argument but part of a fuller vision of reality. Only with this appreciation can we evaluate the atemporal world he describes and so fully understand the reasoning behind, and the consequence of, his assertion that time is unreal.

In Chapter 3 I looked at McTaggart's argument against the reality of time. In §§ 3.2.1-3.2.3 we saw how McTaggart divides time into two series (*A* and *B*) and using this classification, provides an argument for denying its reality. We then saw that accepting this presents us with psychological problems. In § 3.2.4 we found that the experience of change as we know it can no longer be held. In § 3.2.5 we saw that our experience of time in the *B* series must also be wrong. In addition, we also concluded that denying the reality of the *B* series does not provide an alternative to time because still it is denying time as we know it. This makes many arguments against McTaggart's denial, though strongly made, unsuccessful in re-establishing an acceptable temporality. In § 3.3 we saw that if we accept that it is events that make up the nature of change, then any apparent changes in things are misperceptions. However, to adopt this view demands that we hold a strong idealist view and expects us to reject much of what seems part of our experience. In § 3.4 we saw that conventional attacks on the vicious regress are not altogether convincing. This provides strong (if not convincing) support for McTaggart's argument against time. After all, on McTaggart's argument, if the vicious regress is true, time is logically contradictory. I concluded that we have reasonable grounds for accepting atemporality, although common sense tells us otherwise. I acknowledged that it is difficult to accept that it is events and not things that change — we have to take a strong idealist view to accept this and many contemporary philosophers find this difficult — however, defence of the vicious regress can be found.

Working through McTaggart's argument against time in this way, provided us with enough suspicion about the reality of time to encourage us to see what it is he replaces it with. In doing this we took a decisive step forward from nearly ninety years of inconclusive argument about McTaggart's paradox.

In Chapter 4 we saw that none of our commonly held views give us good reason to accept the *A* series. In § 4.2 we saw that although we can take a variety of views about time, none of them is positively conclusive. We saw that the idea of the non-existence of time encourages us to think that it cannot be 'time' that does not exist and whatever is happening that appears as time must be something else. Even so, we found that 'common sense' is something worth holding onto; it is a useful measure against which we can lay the riddles of our experience of temporality. Having looked more closely at the *A* series in § 4.3 (which, based on McTaggart's view is essential for the reality of time) the matter only seemed to get worse. In § 4.3.1 we found that although the concept of past has important social and psychological consequences, as a separate phase of change it shows itself inaccessible to coherent description or proof. In § 4.3.2 we saw that present, like past, fails any attempt at sufficient description. In § 4.3.3. we found that the fact that future cannot be known until it becomes present causes us to suspect that there is indeed any such thing as future, and that we have produced the term only as a means of explaining away the puzzling re-occurrence of 'more' present. I concluded that although the *A* series remains an inescapable part of how time appears to us, analysis of its phases produced various conflicting and unconvincing results. Although it could be that this sort of analysis of the *A* series is misleading; it is possible that time will not yield to this sort of scrutiny and this, by itself, does not mean that it is unreal. But this is improbable. If time in the *A* series is real then its independent phases should be credible. We have seen that this is not so and should not expect to find any greater sense from them if we bundle them together. The *A* series certainly provides us with a convenient everyday mechanism for ordering our lives but its inherent inconsistency does nothing to vouch for its (and consequently time's) reality. The *A* series is not itself an escape route from its own inherent problems nor a consistent system for temporal ordering.

I concluded that it is unsatisfactory that no conventional view provides an answer to some natural questions we have about time's nature. I took the view that on rational and common sense grounds we are justified (like McTaggart) to look outside its terms for an answer.

Even so, there remains strong arguments for believing that time (and our experience of it) can be interpreted sensibly. To guard against taking too rash a step away from time, in Chapter 5, I looked at some positions that we can adopt in relation to our experience of time. In § 5.2, when we looked at the position of 'reality of consciousness plus being', we found strong reasons to believe that McTaggart's denunciation of time should not be dealt with as if it were an end-in-itself. These reasons were strong enough to support withdrawal from the argument over the adequacy or otherwise of the details of McTaggart's argument and its ability to successfully denounce time, and instead move towards a fuller analysis of McTaggart's atemporal replacement. The idea of matching McTaggart's view, first with his fuller concept of the *C* series and then with the modern scientific view of space-time, presented themselves as natural extensions to what we found here. In § 5.3, when we looked at the position 'reality of becoming plus consciousness', we saw that this position requires an acceptance of temporality. It requires that unless there be something in which change can 'happen' there can be no reality of becoming. If we are to accept this position, and the consequent reality of time, we must also accept the confusions that time presents us with. In § 5.4 we looked at the position of 'a four-dimensional universe with no becoming'. We saw that McTaggart's idealist view of the *C* series has similarity to Gödel's world structure. On the face of it this is surprising. McTaggart, as probably the last idealist in the 'British tradition', seems ill-placed in this contemporary world of science. However, we found some similarities between the *C* series and space-time and I concluded that McTaggart's system, because of its complete rejection of temporality, may have greater scope. I concluded that McTaggart's position is surprisingly close to the views held by science. Indeed, it seems that a 'modern idealism' like his has much to offer the scientific model. In § 5.5 we saw that atemporality can indeed be a reasonable view to hold. It also appeals to some of our common sense values — battered as they are by the inherently puzzling nature of time — and that we have no reason to hold back from looking at an atemporal replacement. Because McTaggart proposes such a coherent atemporal system, we are naturally led to investigate his claim for timeless reality.

In Chapter 6 I looked at the replacement for time that McTaggart proposes. In § 6.2 we saw that atemporality allows us to interpret our existence in a way that the puzzling nature of time constantly confounds. It provides a 'replacement' for time and a positive framework for atemporal existence. In doing this, it largely sets aside the temporal puzzle as we experience it. In § 6.3 we saw that the description McTaggart provides of the atemporal *C* series fits in with his fuller description of the nature of existence in general. The point being that the *C* series is part of a complete metaphysical description of reality and is not meant simply to replace time. This, I believe, makes the *C* series a serious contender to fill the gap left by the inadequacy of time to answer implicit problems with temporality. Any (even partial) acceptance of the atemporal system places great strain on any trust we may have that our *B*-experience reflects reality. In § 6.4 we saw that the *C* series can illuminate our ordinary conception of time. It was seen that the 'inclusion series' can be lined up with the asymmetrical nature of time and that its directionality is similar to the apparent direction of the *B* series. In this way the atemporal vision, which is itself 'removed' from time, can be identified with our basic experience of the *B* series and so remain grounded on our unavoidably temporal world. In § 6.5 we saw that time is an inescapable phenomenon. However, I concluded that still we can hold to the notion of the atemporal series and look for fundamental relations between it and our inescapable experience of time. We found that there is a strong argument, particularly in respect of the notion of a common time series, to uphold the acceptability of the real world as atemporal. In § 6.6 I explained how we should alter the common distinction we make between *actual* and *apparent* time. 'Actual' should not be used to denote time which is real as opposed to 'apparent' which is only an impression that is somehow caused by time. Because of the unavoidability of our experience of time, I hold that both 'actual' and 'apparent' can be called 'true'. Calling them 'true', however, does not make them real. If we want to know what really is then we must look behind time to what we can call 'really true' ('time') which is the atemporal *C* series (and, of course, not really time at all). I believe that we need to extend our view of time in this way if we are to understand the *C* series and its relation to the temporal world as we experience it. In § 6.7 I concluded that it is reasonable to look beyond McTaggart's claim that time is unreal. However, I acknowledge that his atemporal world is not easy to accept in total. Indeed, this is at the root of the problem; because McTaggart proposes a complete metaphysics we get easily bound up with his ontological quest. This can lead us to reject the atemporal world for the wrong reasons. To remedy this we need clearly to separate the time of our experience from atemporality which is the reality that lies behind it.

Wittgenstein accuses Freud of being influenced by the nineteenth century view of dynamics.[1] McTaggart was a victim of the same pressure. However, I propose that although his 'dynamic' view demands much us, this does not necessarily lead us to wholesale rejection of his thesis.

Certainly, much of McTaggart's argument against time needs analyzing against his own mystic and psychological motivation. His method, in its apparently clear, fair and logical format, disguises his conviction that ultimately we will all find a place in a heaven of loving individuals. Like Plotinus, McTaggart's mystic vision of idealist reality reaches far beyond the sensations of the world of appearance. As Plotinus sought the fuller truth of Platonism so McTaggart tries to reach the heart of Hegel's Absolute. McTaggart's self passes through a spiritual metempsychosis; from the *fragmentary parts* of time in the apparently material world, by stages of fuller perception, it reaches for the timeless One of reality. It is a system of recurring 'self-rebirth' as new perceptions are continually added to the total content of each self. The final stage is the state of everlasting and timeless lovingness of which all selves will ultimately take part. I agree with Geach, McTaggart makes an 'exemplary effort' to be clear on this subject,[2] but it remains too much McTaggart's 'psychological preoccupation' to be considered on a par with his evaluation of atemporality. Like Nathan (who argues against the reality of McTaggart's self[3]) still I find myself undecided on this 'loving ideality'. The attack he makes on all areas of apparent reality, can cause us to reject his philosophy in its entirety. But we should not fall victim to this. The fact that his system exposes all that we experience to be in error is no reason to doubt the reality of the atemporal world. In his description of the *C*-world we find a coherent replacement for time that does not demand we accept his ontological vision in total.

What we need to do is to isolate the replacement for time from this. In so doing we have a coherent atemporal framework uncomplicated by his fuller mystic vision. What I have tried to do here is just this. I have tried to separate, as far as possible, the atemporal *C*-world from those 'imponderables' (while at the same time duly taking them into account) and present McTaggart's view of atemporality as a meaningful alternative to time.

I believe that this leads us convincingly to the view that the *A* series, in both McTaggart's thinking and in the view of common sense, is untenable. We have found much to support the idea of the *B* series as a series of misperception and little to support it as an underlying fabric for time. We have even discovered that the *C* series can find some support from the empirically-based idea of space-time. In a number of ways we have found that there is good reason to accept the principle of the *C* series. From here we can, if we choose, find that we can regard the existence of our 'selves' with fuller ontological meaning. If we wish to do so, then we have a system against which we can plot a course. On the other hand, we can stop short of this quest. We can accept the atemporal series without becoming embroiled in McTaggart's arguments in defence of the apparent, wholesale error in the universe or the long term conflict between good and evil.

As far as time is concerned, if we wish to keep it separate, the situation is very simple. Everything we experience is unavoidably in time. Although this is a real experience it is not an experience of reality. Even though our experience in time is erroneous it contains a 'reflection' of timeless reality. This 'reflection' provides an acceptable link between our experience in the temporal world and the atemporal world of reality. We do not have to make any 'spiritual commitment' to accept this.

I am convinced that elements of McTaggart's atemporal vision are worthy of further scrutiny in the contemporary world and should not be thrown out just because they form part of a suspicious whole. I think the force of his idealist argument against something that is inherently puzzling (time) has a place in the modern world. It may be that it can contribute to contemporary thinking by resurrecting the metaphysical worth of idealism. I urge that future discussion on McTaggart takes his replacement for time into account.

Notes

1 L. Wittgenstein, *Lectures and Conversations on Aesthetics, Psychology and Religious Belief*, ed. C. Barrett (Berkeley, Calif., University of California Press, 1967), p. 48.

2 Geach, *Truth, Love and Immortality*, p. 165.

3 N.M.L. Nathan, 'McTaggart's Immaterialism', *The Philosophical Quarterly*, vol. 41 (1991), no. 165, pp. 442-56.

Bibliography[1]

Anscombe, G.E.M. 'Aristotle and the Sea Battle, De Interpretatione, Ch. IX', in Smart, *Problems of Space and Time*, pp. 43-57. A revised version, with some omissions and additions, originally in *Mind*, vol. 65 (1956), no. 257, pp. 1-15.

Aristotle, *The Basic Works of Aristotle*, ed. and with an Introduction by R. McKeon (New York, Random House, 1941).

— *De Interpretatione (On Interpretation)*, trans. by E.M. Edghill, in McKeon, *The Basic Works of Aristotle*, pp. 38-61.

— *Physica (Physics)*, trans. by R.P. Hardie and R.K. Gaye, in McKeon, *The Basic Works of Aristotle*, pp. 213-394.

— *De Anima (On the Soul)*, trans. by J.A. Smith, in McKeon, *The Basic Works of Aristotle*, pp. 533-603.

— *Metaphysica (Metaphysics)*, trans. by W.D. Ross, in McKeon, *The Basic Works of Aristotle*, pp. 681-926.

— *De Poetica* (*Poetics*), trans. by I. Bywater, in McKeon, *The Basic Works of Aristotle*, pp. 1453-87.

Augustine, St. *Confessions*, ed. W.H.D. Rouse, trans. by W. Watts (London, Heinemann, 2 vols., 1912).

— *City of God*, trans. by J. Healey, Introduction by E. Barker (London, J.M. Dent, 1931).

Ayer, A.J. *Philosophical Essays* (London, Macmillan, 1954).

Bergson, H. *Time and Free-will*, trans. by F.L. Pogson (London, Allen and Unwin, 1910).

— *Creative Evolution*, trans. by A. Mitchell (London, Macmillan, 1911).

Berkeley, G. *De Motu* in *Philosophical Works including the Works on Vision*, eds. A.A. Luce and T.E. Jessop, Introduction and Notes by M.R. Ayers, (London, J.M. Dent, 1985).

Biglow, J. 'Worlds Enough for Time', *Noûs*, vol. 25 (1991), pp. 1-19.

Broad, C.D. *Scientific Thought* (London, Kegan Paul, Trench, Trubner and Co., 1923).

— *Examination of McTaggart's Philosophy* (Cambridge University Press, 2 vols., vol. 1, 1933; vol. 2 (Parts 1 and 2), 1938).

— 'The Local Historical Background of Contemporary Cambridge Philosophy', in C.A. Mace, ed., *British Philosophy in the Mid-Century* (London, George Allen and Unwin, 1957).

Buller, D.J. and Foster, T.R. 'The New Paradox of Temporal Transcience', *The Philosophical Quarterly*, vol. 42 (1992), no. 168, pp. 357-66.

Cajori, F. *Newton's Principia*, trans. by A. Motte (Berkeley, Calif., California University Press, 1934).

Cedarholm, J.P. and Townes, C.H. 'A New Experimental Test of Special Relativity', *Nature* (1959), no. 184, pp. 1350-1.

Christensen, F. 'McTaggart's Paradox and the Nature of Time', *The Philosophical Quarterly*, vol. 24 (1974), no. 97, pp. 289-99.

Cleugh, M.F. *Time and its Importance in Modern Thought* (London, Methuen, 1937).

Craig, W.L. 'Tachyons, Time Travel, and Divine Omniscience', *Journal of Philosophy*, vol. 85 (1988), no. 3, pp. 135-50.

Currie, G. 'McTaggart at the Movies', *Philosophy*, vol. 67 (1992), no. 261, pp. 343-55.

Dalton, J. *A New System of Chemical Philosophy* (Manchester University Press, 2nd edn., 1808).

Davies, P. 'Time Asymmetry and Quantum Mechanics', in Flood and Lockwood, *The Nature of Time*, pp. 99-124.

Deutsch, D. *The Fabric of Reality* (Harmondsworth, Penguin Press, 1997).

Dummett, M. 'A Defence of McTaggart's Proof of the Unreality of Time', *Philosophical Review*, vol. 69 (1960), no. 392, pp. 497-504.

— 'Causal Loops', in Flood and Lockwood, *The Nature of Time*, pp. 135-69.

Dunne, J.W. *An Experiment with Time* (London, Faber and Faber, 3rd edn., 1934. First published 1927).

— *The Serial Universe* (London, Faber and Faber, 1934).

Earman, J. 'An Attempt to Add a Little Direction to "The Problem of the Direction of Time"', *Philosophy of Science*, vol. 41 (1974), pp. 15-47.

— *World Enough and Space-Time* (Cambridge, Mass., MIT Press, 1989).

Earman, J., Glymour, C. and Stachel, J. eds., *Foundations of Space-Time Theories, Minnesota Studies in the Philosophy of Science*, vol. 8 (Minneapolis, University of Minnesota Press, 1977).

Eddington, E.S. *The Nature of the Physical World*, Gifford Lectures, 1927 (Cambridge University Press, 1928).

Einstein, A. *Relativity, the Special and the General Theory: A Popular Exposition*, trans. by R.W. Lawson (London, Methuen, 1988. First published 1920).

Evans-Pritchard, E.E. *Witchcraft, Oracles and Magic among the Azande* (Oxford, Clarendon Press, 1937).

Farmer, D.J. *Being in Time: The Nature of Time in Light of McTaggart's Paradox* (Lanham, Maryland, University Press of America, 1990).

Findlay, J.N. 'Time: A Treatment of Some Puzzles', in A.G.N. Flew, ed., *Essays on Logic and Language* (Oxford, Basil Blackwell, 1951), pp. 37-54.

Flood, R. and Lockwood, M. eds., *The Nature of Time* (Oxford, Basil Blackwell, 1988).

Gale, R.M. 'Endorsing Predictions', *The Philosophical Review*, vol. 70 (1961), no. 395, pp. 376-85.

— 'Tensed Statements', *The Philosophical Quarterly*, vol. 12 (1962), no. 46, pp. 53-9.

— 'A Reply to Smart, Mayo and Thalberg on "Tensed Statements"', *The Philosophical Quarterly*, vol. 13 (1963), no. 53, pp. 351-6.
— 'Is it Now Now?', *Mind*, vol. 73 (1964), no. 289, pp. 97-105.
— 'Existence, Tense, and Presupposition', *The Monist*, vol. 50 (1966), no. 1, pp. 98-108.
— 'McTaggart's Analysis of Time', *American Philosophical Quarterly*, vol. 3 (1966), no. 2, pp. 145-52.
— *The Language of Time* (London, Routledge and Kegan Paul, 1968).
— ed. *The Philosophy of Time* (London, Macmillan, 1968).
— 'The Static Versus the Dynamic Temporal', in Gale, *The Philosophy of Time*, pp. 65-85.
Geach, P.T. *Truth, Love and Immortality, An Introduction to McTaggart's Philosophy* (London, Hutchinson, 1979).
Glymour, C. 'Topology, Cosmology and Convention', *Synthèse*, vol. 24 (1972), nos. 1/2, pp. 195-218.
— 'Indistinguishable Space-Times and the Fundamental Group', in Earman, Glymour and Stachel, *Foundations of Space-Time Theories*, pp. 50-60.
Gödel, K. 'A Remark about the Relationship between Relativity Theory and Idealistic Philosophy', in Yourgrau, *Demonstratives*, pp. 261-5. Originally in Schilpp, *Albert Einstein — Philosopher-Scientist, Library of Living Philosophers*, pp. 557-62.
Gotshalk, D.W. 'McTaggart on Time', *Mind*, vol. 39 (1930), no. 153, pp. 26-42.
Grünbaum, A. *Philosophical Problems of Space and Time* (New York, Knopf, 1963).
— 'The Status of Temporal Becoming', in Gale, *The Philosophy of Time*, pp. 322-54.
— 'The Meaning of Time', in N. Rescher et al., eds., *Essays in Honor of Carl G. Hempel* (Dordrecht, 1969), pp. 147-77.
Gunn, J.A. *The Problem of Time, An Historical and Critical Study* (London, George Allen and Unwin, 1929).
Harrison, J. 'Dr Who and the Philosophers', *Proceedings of the Aristotelian Society*, vol. 45 (1971), pp. 1-24.
Hawking, S. *A Brief History of Time* (London, Bantam Press, 1988).
Heidegger, M. *Being and Time*, trans. by J. Macquarrie and E. Robinson (New York, Harper and Row, 1962).
Heller, M. 'Things Change', *Philosophy and Phenomenological Research*, vol. 52 (1992), pp. 695-704.
Hollis, M. 'Times and Spaces', *Mind*, vol. 76 (1967), no. 304, pp. 524-36.
Homer, *The Iliad*, trans. by E.V. Rieu (Harmondsworth, Penguin, 1985).
Horwich, P. 'On Some Alleged Paradoxes of Time Travel', *Journal of Philosophy*, vol. 72 (1975), no. 14, pp. 432-44.
Hume, D. *A Treatise of Human Nature*, ed. with an Analytical Index by L.A. Selby-Bigge, 2nd edn., revised by P.H. Nidditch (Oxford University Press, 1978).
Husserl, E. *The Phenomenology of Internal Time-Consciousness*, trans. by J.S. Churchill (Bloomington, Indiana University Press, 1964).

James, W. 'The Perception of Time', *The Principles of Psychology*, ch. 15 (Chicago, Encyclopaedia Britannica, 1952. First published 1891), pp. 396-420.
— 'Memory', *The Principles of Psychology*, ch. 16 (Chicago, Encyclopaedia Britannica, 1952. First published 1891), pp. 421-51.
Kant, I. *Critique of Pure Reason*, (1781; 2nd edn., 1787), trans. by N. Kemp Smith (London, Macmillan, 1973).
Kaplan, D. 'Thoughts on Demonstratives', in Yourgrau, *Demonstratives*, pp. 34-49.
Keeling, S.V. Note, 'Request for MS', *Mind*, vol. 42 (1933), no. 168, p. 544.
— Note, 'McTaggart's [*The*] *Nature of Existence*, vol. I, Comments and Amendments', *Mind*, vol. 47 (1938), no. 188, pp. 547-50.
— *Time and Duration*, ed. G. Rochelle with an Introduction by E. Senior (Lewiston, New York, Edwin Mellen, 1991).
Kiernan-Lewis, D. 'Not Over Yet: Prior's "Thank Goodness" Argument', *Philosophy*, vol. 66 (1991), no. 256, pp. 241-3.
— 'The Rediscovery of Tense: A Reply to Oaklander', *Philosophy*, vol. 69 (1994), no. 268, pp. 231-3.
King-Farlow, J. 'The Positive McTaggart on Time', *Philosophy*, vol. 49 (1974), no. 188, pp. 169-78.
Kochler, W. *The Mentality of Apes*, trans. from the 2nd revised edn. by E. Winter (Harmondsworth, Penguin, 1957).
Le Poidevin, R. *Change, Cause and Contradiction: A Defence of the Tenseless Theory of Time* (London, Macmillan, 1991).
— Book Review, 'D.J. Farmer, *Being in Time: The Nature of Time in Light of McTaggart's Paradox*', *Mind*, vol. 100 (1991), no. 3, pp. 388-90.
— 'Lowe on McTaggart', *Mind*, vol. 102 (1993), no. 405, pp. 163-70.
Le Poidevin, R. and MacBeath, M. eds., *The Philosophy of Time* (Oxford University Press, 1993).
Le Poidevin, R. and Mellor, D.H. 'Time, Change, and the "Indexical Fallacy"', *Mind*, vol. 96 (1987), no. 384, pp. 534-8.
Leibniz, G.W. *The Monadology*, trans. by R. Latta (Oxford, Clarendon Press, 1925).
— *Philosophical Writings*, ed. and Introduction by G.H.R. Parkinson, trans. by M. Morris and G.H.R. Parkinson (London, J.M. Dent, 1988).
Loizou, A. *The Reality of Time* (Aldershot, Gower, 1986).
Lombard, L.B. 'The Doctrine of Temporal Parts and the "No-Change" Objection', *Philosophy and Phenomenological Research*, vol. 54 (1994), pp. 365-72.
Lowe, E.J. 'The Indexical Fallacy in McTaggart's Proof of the Unreality of Time', *Mind*, vol. 96 (1987), no. 381, p. 62-70.
— 'Reply to Le Poidevin and Mellor', *Mind*, vol. 96 (1987), no. 384, pp. 539-42.
— 'McTaggart's Paradox Revisited', *Mind*, vol. 101 (1992), no. 402, pp. 323-6.
— 'Comment on Le Poidevin', *Mind*, vol. 102 (1993), no. 405, pp. 171-3.
Lucas, J.R. *A Treatise on Time and Space* (London, Methuen, 1973).
— 'The Open Future', in Flood and Lockwood, *The Nature of Time*, pp. 125-34.
— *The Future* (Oxford, Basil Blackwell, 1989).

Luce, J.V. *An Introduction to Greek Philosophy* (London, Thames and Hudson, 1992).

Lucretius, 'De rerum natura', Bk 1. in *Lucretius: on the Nature of the Universe*, Introduction and trans. by R.E. Latham (Harmondsworth, Penguin, 1951).

MacBeath, M. 'Who was Dr Who's Father?', *Synthèse*, vol. 51 (1982), no. 3, pp. 397-430.

— 'Communication in Reversed Time', *Synthèse*, vol. 56 (1983), no. 1, pp. 27-46.

— 'Clipping Time's Wings', *Mind*, vol. 95 (1986), no. 378, pp. 233-7.

— 'Time's Square', in Le Poidevin and MacBeath, *The Philosophy of Time*, pp. 183-202.

Mach, E. *Science of Mechanics*, trans. by T.J. McCormack, 6th edn. (Chicago, Open Court, 1960).

McTaggart, J.McT.E. 'The Further Determination of the Absolute'. Privately published 1893. Reproduced in *Philosophical Studies*, pp. 210-72.

— 'The Necessity of Dogma', *International Journal of Ethics*, vol. 5 (1895), no. 2, pp. 147-62.

— *Studies in the Hegelian Dialectic* (Cambridge University Press, 1896, 2nd edn., 1922), 255 pp.

— 'Introduction to the Study of Philosophy' (McTaggart's personal notes for the course of Friday evening lectures held at Trinity College, Cambridge, from 1899-1914). Reproduced in *Philosophical Studies*, pp. 183-209.

— *Studies in Hegelian Cosmology* (Cambridge University Press, 1901, 2nd edn., 1918), 293 pp.

— 'Some Considerations Relating to Human Immortality', *International Journal of Ethics*, vol. 13 (1903), no. 2, pp. 152-71.

— 'Human Pre-existence', *International Journal of Ethics*, vol. 15 (1904), no. 1, pp. 83-95.

— *Some Dogmas of Religion*, 2nd edn., ed. and with Introduction by C.D. Broad, including McTaggart's list of emendations extracted by S.V. Keeling, new Introduction by G. Rochelle (Bristol, Thoemmes Press, 1997. First published London, Edward Arnold, 1906; 2nd edn., 1930). 299 pp.

— 'The Relation of Time and Eternity', an address before the Philosophical Union of the University of California, 23 Aug. 1907. *University of California Chronicle*, vol. 10 (1908), no. 2, pp. 127-52. Reprinted as a pamphlet (Berkeley, Calif., The University Press, 1908), 28 pp. Reprinted in *Mind*, vol. 18 (1909), no. 71, pp. 343-62. Reproduced in *Philosophical Studies*, pp. 132-55.

— 'The Individualism of Value', *International Journal of Ethics*, vol. 18 (1908), no. 4, pp. 433-45. Reproduced in *Philosophical Studies*, pp. 97-109.

— 'The Unreality of Time', *Mind*, vol. 17 (1908), no. 68, pp. 457-74. Reproduced in *Philosophical Studies*, pp. 110-31.

— 'Mysticism', *The New Quarterly*, vol. 2 (1909), no. 7, pp. 315-39. Reproduced in *Philosophical Studies*, pp. 46-68.

— *Dare to be Wise* (London, Watts and Co. (for The Rationalist Press Association), 1910), 9pp. Reproduced in *Philosophical Studies*, pp. 37-45.

— *Commentary on Hegel's Logic* (Cambridge University Press, 1910), 311 pp.
— *Human Immortality and Pre-Existence* (London, Edward Arnold, 1915), 119 pp. Reprint of Chapter 3 and Chapter 4 of *Some Dogmas of Religion.*
— 'The Meaning of Causality', *Mind*, vol. 24 (1915), no. 95, pp. 326-44. Reproduced in *Philosophical Studies*, pp. 156-78.
— 'Personality', *Encyclopaedia of Religion and Ethics*, vol. 9, ed. J. Hastings (Edinburgh, T. and T. Clark, 1917), pp. 773-781. Reproduced in *Philosophical Studies*, pp. 69-96.
— *The Nature of Existence* (Cambridge University Press, 2 vols., vol. 1, 1921), 309 pp., reprinted 1968 and 1988, 330 pp; vol. 2, ed. C.D. Broad, 1927, 479 pp., reprinted 1968 and 1988, 526 pp.
— 'Propositions Applicable to Themselves', *Mind*, vol. 32 (1923), no. 128, pp. 462-4. Reproduced in *Philosophical Studies*, pp. 179-82.
— 'An Ontological Idealism', *Contemporary British Philosophy*, 1st Series, ed. J.H. Muirhead (London, George Allen and Unwin, 1924), pp. 251-69. Reproduced in *Philosophical Studies*, pp. 273-92.
— *Philosophical Studies*, ed. and Introduction by S.V. Keeling, Foreword by Gerald Rochelle (Bristol, Thoemmes Press, 1996. First published, London, Edward Arnold, 1934). 292 pp.[2]
Malament, D. 'Observationally Indistinguishable Space-times: Comments on Glymour's Paper', in Earman, Glymour and Stachel, *Foundations of Space-Time Theories*, pp. 61-80.
Mander, W. J. 'On McTaggart on Love', *History of Philosophy Quarterly*, vol. 13 (1996), no. 1, pp. 133-47.
Marhenke, P. 'McTaggart's Analysis of Time', *University of California Publications in Philosophy*, vol. 18 (1935), pp. 151-73.
Markosian, N. 'How Fast Does Time Pass?', *Philosophy and Phenomenological Research*, vol. 53 (1993), pp. 829-44.
Maxwell, J.C. *A Treatise on Electricity and Magnetism* (Oxford, Clarendon Press, 2 vols., 1873).
Meiland, J.W. 'A Two-Dimensional Passage Model of Time for Time Travel', *Philosophical Studies*, vol. 26 (1974), nos. 3/4, pp. 153-73.
Mellor, D.H. *Real Time* (Cambridge University Press, 1987. First published 1981).
— 'Fixed Past, Unfixed Future', in B. Taylor, ed. *Michael Dummett: Contributions to Philosophy* (Dordrecht, Nijhoff, 1987), pp. 168-88.
— 'Causation and the Direction of Time', *Erkenntnis*, vol. 35 (1991), nos. 1/3, pp. 191-203.
Merleau-Ponty, M. *The Phenomelogy of Perception*, trans. by C. Smith (London, Routledge and Kegan Paul, 1962).
Michelson, A.A. 'On the Relative Motion of the Earth and the Luminiferous Ether', *The American Journal of Science and Arts*, vol. 22 (1881), no. 128, pp. 120-9.
Mink, L.O. 'Time, McTaggart and Pickwickian Language', *The Philosophical Quarterly*, vol. 10 (1960), no. 40, pp. 252-63.

Minkowski, H. 'Space and Time', in H.A. Lorentz, A. Einstein, H. Minkowski, and H. Weyl, *The Principle of Relativity. A Collection of Original Memoirs on Special and General Theory of Relativity*, with notes by A. Sommerfeld, trans. by W. Perrett and G.B. Jeffery (New York, Dover, 1952. First published 1923), pp. 75-91.

Misner, C.S., Thorne, K.S. and Wheeler, A., *Gravitation* (San Francisco, W.H. Freeman, 1973).

Nathan, N.M.L. 'McTaggart's Immaterialism', *The Philosophical Quarterly*, vol. 41 (1991), no. 165, pp. 442-56.

Newton-Smith, W.H. *The Structure of Time* (London, Routledge and Kegan Paul, 1980).

Nusenoff, R.E. 'Spacialized Time Again', *Philosophy*, vol. 52 (1977), no. 199, pp.100-1.

— 'Two-Dimensional Time', *Philosophical Studies*, vol. 29 (1976), no. 5, pp. 337-41.

Oakeley, H.D. 'Time and the Self in McTaggart's System', *Mind*, vol. 39 (1930), no. 154, pp. 175-93.

Pears, D.F. 'Time, Truth and Inference', *Proceedings of the Aristotelian Society*, vol. 51 (1950-51), pp. 1-24.

Penrose, R. *The Emperor's New Mind* (Oxford University Press, 1989).

Piaget, J. *The Child's Conception of Time*, trans. by A.J. Pomerans (London, Routledge and Kegan Paul, 1969).

Piaget, J. and Inhelder, B. *The Child's Conception of Space*, trans. by F.J. Langdon and J.L. Lunzer (London, Routledge and Kegan Paul, 1956).

Plato, *The Republic*, in *The Dialogues of Plato*, with Analyses and Introductions and trans. by B. Jowett (Oxford, Clarendon Press, 1875), pp. 1-519.

Plumer, G. 'Why Time is Extensive', *Mind*, vol. 93 (1984), no. 370, pp. 265-70.

Powers, J. *Philosophy and the New Physics* (London, Methuen, 1982).

Price, H.H. 'The Appeal to Common Sense', *Journal of Philosophical Studies*, vol. 5 (1930), no. 17, pp. 24-35.

Priestley, J.B. *Man and Time* (London, W.H. Allen, 1978. First published 1964).

Prior, A.N. 'Thank Goodness That's Over', *Philosophy*, vol. 34 (1959), no. 128, pp. 12-17.

— 'Postulates for Tense-Logic', *American Philosophical Quarterly*, vol. 3 (1966), no. 2, pp. 153-61.

— *Past, Present and Future* (Oxford, Clarendon Press, 1967).

— 'Changes in Events and Changes in Things', in Le Poidevin and MacBeath, *The Philosophy of Time*, pp. 35-46.

Putnam, H. 'Time and Physical Geometry', *Journal of Philosophy*, vol. 64 (1967), pp. 240-7.

— *Reason, Truth and History* (Cambridge University Press, 1981).

Quine, W.V. *Ways of Paradox* (New York, Random House, 1966).

Quinton, A. 'Spaces and times', *Philosophy*, vol. 37 (1962), no. 140, pp. 130-47.

— *The Nature of Things* (London, Routledge and Kegan Paul, 1973).

Radin, P. *Primitive Man as Philosopher* (New York, Dover, 1957).
Ray, C. *Time, Space and Philosophy* (London, Routledge and Kegan Paul, 1991).
Reichenbach, H. *Elements of Symbolic Logic* (New York, Macmillan, 1947).
— *The Direction of Time* (Berkeley, Calif., University of California Press, 1956).
— *The Philosophy of Space and Time*, trans. by M. Reichenbach and J. Freund (New York, Dover, 1958).
Rietdijk, C. 'A Rigorous Proof of Determinism derived from the Special Theory of Relativity', *Philosophy of Science*, vol. 33 (1966), pp. 341-4.
Riggs, P.J. 'A Critique of Mellor's Argument against "Backwards" Causation', *British Journal of Philosophy*, vol. 42 (1991), pp. 75-86.
Rochelle, G. *The Life and Philosophy of J.McT.E. McTaggart, 1866-1925* (Lewiston, New York, Edwin Mellen, 1991).
Rosen, S. 'Thought and Touch. A Note on Aristotle's *De Anima*', in Yourgrau, *Demonstratives*, pp. 185-94.
Russell, B. 'On the Experience of Time', *The Monist*, vol. 25 (1915), no. 2, pp. 212-33.
— *Mysticism and Logic* (London, George Allen and Unwin, 1918).
— *Human Knowledge. Its Scope and Limits* (London, George Allen and Unwin, 1948).
— *The Principles of Mathematics*, 2nd edn. (London, George Allen and Unwin, 1985. First published 1903).
Ryle, G. *The Concept of Mind* (Harmondsworth, Peregrine, 1983. First published, 1949).
Sandford, D. 'McTaggart on Time', *Philosophy*, vol. 43 (1968), no. 166, pp. 371-8.
Schilpp, P.A. ed. *Albert Einstein — Philosopher-Scientist, Library of Living Philosophers*, vol. 7 (La Salle, Ill., Open Court, 1949).
Schlesinger, G.N. 'How Time Flies', *Mind*, vol. 91 (1982), no. 364, pp. 501-23.
— 'E Pur Si Muove', *The Philosophical Quarterly*, vol. 41 (1991), no. 165, pp. 427-41.
Sciama, D. 'Time "Paradoxes" in Relativity', in Flood and Lockwood, *The Nature of Time*, pp. 6-21.
Shallis, M. 'Time and Cosmology', in Flood and Lockwood, *The Nature of Time*, pp. 63-79.
Shea, W.R. 'McTaggart and the Neo-positivist Entropists', *Philosophy*, vol. 50 (1975), no. 193, pp. 346-51.
Shoemaker, S. 'Time Without Change', *Journal of Philosophy*, vol. 66 (1969), no. 12, pp. 363-81.
Sklar, L. *Space, Time and Spacetime* (Berkeley, Calif., University of California Press, 1977).
— 'What Might be Right about the Causal Theory of Time', *Synthèse*, vol. 35 (1977), no. 2, pp. 155-71.
— 'Up and Down, Left and Right, Past and Future', in Le Poidevin and MacBeath, *The Philosophy of Time*, pp. 99-116. First published in *Noûs*, vol. 15 (1981), no. 2, pp. 111-29.

— 'Time, Reality, and Relativity', in Yourgrau, *Demonstratives*, pp. 247-60.
Smart, J.J.C. 'The River of Time', *Mind*, vol. 58 (1949), no. 232, pp. 483-94.
— *Philosophy and Scientific Realism* (London, Routledge and Kegan Paul, 1963).
— ed. *Problems of Space and Time* (New York, Macmillan, 1964).
— 'Time and Becoming', in *Time and Cause: Essays Presented to Richard Taylor*, ed. P. van Inwagen (Dordrecht, Reidel, 1980), pp. 3-15.
Spinoza, B. *Ethics*, revised by G.H.R. Parkinson, trans. by A. Boyle (London, J.M. Dent, 1985).
Stebbing, L.S. 'Some Ambiguities in Discussions Concerning Time', *Philosophy in History: Essays Presented to Ernst Carrirer*, ed. R. Klibansky and H.J. Paton (Oxford, Clarendon Press, 1936), pp. 107-23.
— Critical Notice, 'C.D. Broad, *Examination of McTaggart's Philosophy*, vol. II', *Mind*, vol. 48 (1939), no. 190, pp. 214-20.
Swinburne, R. 'Times', *Analysis*, vol. 25 (1965), no. 108 (NS), pp. 185-91.
— *Space and Time* (London, Macmillan, 1981. First published 1968).
Teichmann, R. 'Time and Change', *The Philosophical Quarterly*, vol. 43 (1993), no. 171, pp. 158-77.
Thomson, J.J. 'Time, Space and Objects', *Mind*, vol. 74 (1965), no. 293, pp. 1-27.
Thucydides, *The History of the Peloponnesian War*, ed. and trans. by R. Livingstone (Oxford University Press, 1960).
Walker, S. *Animal Thought* (London, Routledge and Kegan Paul, 1983).
Welby, V. 'Mr McTaggart on the "Unreality of Time"', *Mind*, vol. 18 (1909), no. 70, pp. 326-8.
Whitrow, G.J. *The Natural Philosophy of Time* (Oxford, Clarendon Press, 1980).
— *Time in History* (Oxford University Press, 1989).
Whorf, B.L. *Language, Thought and Reality*, ed. J.B. Carroll (Cambridge, Mass., MIT Press, 1956).
Wilkerson, T.E. 'Time and Time Again', *Philosophy*, vol. 48 (1973), no. 184, pp. 173-7.
— 'More Time and Time Again', *Philosophy*, vol. 54 (1979), no. 207, pp. 110-12.
Williams, D.C. 'The Myth of Passage', *Journal of Philosophy*, vol. 48 (1951), no. 15, pp. 457-72.
Wisdom, J. 'McTaggart's Determining Correspondence of Substance: A Refutation', *Mind*, vol. 37 (1928), no. 148, pp. 414-38.
— 'Time, Fact and Substance', *Proceedings of the Aristotelian Society*, vol. 29 (1928-29), pp. 69-94.
Wittgenstein, L. *Tractatus Logico-Philosophicus* (London, Routledge and Kegan Paul, 1986. First published 1922).
— *Lectures and Conversations on Aesthetics, Psychology and Religious Belief*, ed. C. Barrett (Berkeley, Calif., University of California Press, 1967).
Yourgrau, P. ed. *Demonstratives* (Oxford University Press, 1990).
— *The Disappearance of Time: Kurt Gödel and the Idealistic Tradition in Philosophy* (Cambridge University Press, 1991).

MSS and TSS, *The Nature of Existence* vol. 2: Draft *A* and Notes for Draft *B*; Part of Draft *B*; Part of Draft *C*; The Wren Library, Trinity College, Cambridge.
Letter, Geach, P.T. to Rochelle, G., 6 September 1990.

Notes

1 For a complete list of McTaggart's works, source materials (including private papers) and a bibliography of commentators and critics see, Rochelle, *The Life and Philosophy of J.McT.E. McTaggart 1866-1925*.

2 During his compilation of *Philosophical Studies*, Keeling made an enquiry regarding the whereabouts of the MS of 'The Empirical Elements of Metaphysics', the last address McTaggart delivered to the Moral Science Club at Trinity in 1923 or 1924 (S.V. Keeling, Note, 'Request for MS', *Mind*, vol. 42 (1933), no. 168, p. 544). The enquiry met with no success.

Index